Rattling The Establishment's Cage:

Short Essays & Notes on Unorthodox Views of Environment, Economics, Ethics, Politics, War, Spirituality, and More

by Scott C. Haley

Copyright © 2024 Scott C. Haley

Author's current professional association memberships: Society for Conservation Biology; International Society for Ecological Economics; American Association of Geographers; International Society for Environmental Ethics.

....

Note to Reader

Many of these essays were written for a blog I've maintained since 2007, IndividualSovereignty.blogspot.com; others were penned this year, 2024, and are not on the blog as of this writing. All have been written by me (no A.I.), are some of the most popular on the blog, and all were edited for this book. Each essay is on its own page (some are two or three pages in length), and for context, I've included the year it was written. Enjoy, and

Happy Trails,
Scott Haley

Dedication

To the memory of four of my teachers---

From Highland Park High School, Highland Park, Illinois, 1956-1960:
1. Mr. Ralph Cianchetti, an English teacher who inspired my love of the written word, and encouraged me to find my own "Lake Isle of Innisfree".
2. Mr. Benson, who inspired my love and appreciation of both American Government and Modern European History.

From Arizona State University, Tempe, Arizona, 1971-1973:
1. Dr. Ira Judd, who increased my appreciation of Plant Ecology by leaps and bounds.
2. Dr. Virgil Baker, who had a grasp of Human Ecology which lit a fire in me that still burns brightly.

All were magnificent teachers.

Table of Contents

Environment

August, 2022

Hypothesis: It's likely the Powers-That-Be (public & private) in our Land have no intention whatsoever of...

I've studied human impact on the environment for fifty years, *and watched our "transition" to renewable energy for a tad over twenty years*. We've made a little progress in renewable energy, but fossil fuels still rule the roost. Except for a reduction in the use of coal, and except for oil cartels reducing production once in awhile to manipulate the price, and except for lower demand for gasoline during the first two years of the pandemic, *fossil fuel use has not been significantly reduced in the USA*. Plus, renewable energy production has not been able to meet even the annual *increase (the increase only)* in energy demand here and worldwide.
https://www.cnbc.com/2021/11/04/gap-between-renewable-energy-and-power-demand-oil-gas-coal.html

So, I propose the following hypothesis. "It's likely the Powers-That-Be (public & private) in our Land **have no intention whatsoever** of transitioning to renewable energy. ***Instead,*** the plan is to kick the can down the road while having renewables *complement* the dominant fossil fuels". [Of course, sooner or later, fossil energy either will be too inaccessible to extract affordably, or it will be gone. But that won't be anytime soon, or soon enough.] Meanwhile, a bunch of politicians (funded by the short-sighted fossil industry) have assured the proverbial kicking of the can down the road with the recent Inflation Reduction Act. Said law is somewhat good for the *complementary* renewable energy industry, and even better for the myopic fossil fuel cartel.

I imagine others have come up with this hypothesis before me, but I suspect most of Main Street still believe the elites are sincere when they claim to be transitioning away from fossil fuels to renewables. Thus I fear pressure in favor of accelerating the transition will abate. Such would be extremely detrimental to *Homo sapiens* and all other life on Spaceship Earth.

December, 2022

COP 27 more or less confirms my hypothesis

COP 27, like all its predecessors, failed to endorse the phasing out of all fossil fuels. They did call for phasing out coal, but not oil and natural gas. **Why?** The answer is simple: oil and gas almost are the perfect energy sources to enable the pursuit of infinite, ever-expanding, economic growth. Compared to renewables, they are transportable, more convenient, cheaper to obtain, packed with more punch, and the infrastructure for extraction and distribution already is developed. As to any problems created by fossil fuel use, no worries, new technology will solve them. At least, that's most likely how the linear thinking, neoliberal petroleum supporters view the situation.

Other people see things through the lens of comprehensive thinking and complexity theory. Their conclusions are quite different, and they've been warning us about our current state **for decades**. Plus, the freight train which will hit us relatively soon *involves much more than only climate disruption*.

Either we will significantly address the impending catastrophes of appalling air-water-soil pollution, soil loss, natural ecosystem destruction, overharvesting of forests and fish, the toxic forever-chemicals, nuclear wastes, polluting-destructive-perpetual war, biodiversity loss, the spread of disease, and more, or Mother Nature will. She already has started.

It's beginning to appear that positive action by individual countries on the scale required is not going to happen. Therefore, it's up to smaller political entities - states/provinces, counties, local cities and towns - to adopt and implement sustainable policies **ASAP**. At least then, some areas will have a somewhat better chance of weathering the coming disasters. Which disasters? Things like severe food shortages, a lack of fresh water, a lack of jobs, a sense of insecurity, the feeling of living in an atomized society, a lack of Nature, etc. Much is being done in some areas, and much more can be done everywhere. <u>Search out solutions</u>; they're all over the internet. We're in ecological overshoot already; the human race is in dire straits.

December, 2023

U. S. Government's Lip Service to Climate Mitigation, plus the "Planet Wreckers Report"

According to the Center for Biological Diversity, the Biden Administration's actions re a new round of oil and gas leasing auctions could potentially "erase" any climate gains in the Inflation Reduction Act. The USA is now the world's largest oil producer, and in 2022 **66% of our oil came from the fracking of shale**. Often called shale oil, it's also referred to as "tight oil". As you may know, drilling for it often has resulted in big problems (all of which usually are denied by the industry).

At the global level, we now have twenty countries (including the USA) planning to expand oil production for *twenty-six more years*. https://priceofoil.org/2023/09/12/planet-wreckers-how-20-countries-oil-and-gas-extraction-plans-risk-locking-in-climate-chaos/

In addition, COP 28 was run by the CEO of one of the largest oil producers in the world. Do you think there's any conflict of interest in such a scenario? Seems likely, eh?

As pointed out previously, we do need to use fossil fuels in the immediate future in order to construct and distribute massive numbers of Green energy gizmos and new electrical power lines. I don't know for how long, **but surely not twenty-six more years**.

Is it not yet clear that The Establishment is on the wrong path? And not yet clear that voters keep electing politicians to high office who favor consumerism over ecological principles? If this whole business weren't so tragic, it would be laughable.

July, 2020

What is Mother Earth Telling Us?

As we humans continue deforestation in tropical areas, massive air pollution in temperate regions, and numerous other ecocide activities all over the world, our planet (which is a dynamic ecosystem) is reacting in various ways. The reactions in total are sending a message which is clear to anyone with either an academic ecological background, or anyone with an intuitive sense of ecology, or anyone with both.

Professor William E. Rees deftly explains it here: https://thetyee.ca/Analysis/2020/04/06/The-Earth-Is-Telling-Us-We-Must-Rethink-Our-Growth-Society/

Yes, humanity has done many amazing, wonderful, and productive things in the past 5,000 years or so. There's no denying that. Plus, in approximately the last 150-200 years, we've made advances in technology, health, food production, and more which are nothing short of miraculous.

Unfortunately, however, for the last 50 years or thereabouts, we also have been on a path of species suicide. We have been adhering to an ideology which preaches unlimited growth and overconsumption. For all practical purposes, it's the largest secular religion in developed countries.

The disease, cancer, provides an instructive lesson. What is the overriding characteristic of cancer? It's *unlimited growth.* That occurs in a finite body, and eventually kills the body. In a certain sense, that's what some of our activities are doing to our habitat; and those particular activities are the result of the economic religion to which we are adhering.

The article cited above is dated April 6, 2020. It deals with current problems, and *suggested solutions*. It broaches subjects about which too many of us have had our heads in the sand for too long. If we keep treating the *symptoms* of ecological problems rather than the root causes, organized human existence soon will be in serious jeopardy.

May, 2020

Myopia and Deep Denial by Too Many People

Any informed, reasonable assessment of the current state of humanity on finite Earth surely would conclude with the following.
1. The Powers-That-Be and too many Main Street people apparently believe we humans are separate from and superior to Nature... and thus not subject to the Laws of Nature.
2. As a consequence, we have been and continue to be on a path of subduing Nature, rather than extracting natural capital in a sustainable manner.
3. The results have been: overuse, degradation, and dangerous disruption of our natural habitat (plus, pollution in the extreme).
4. Much of the above is due to propaganda, a lack of knowledge concerning natural science (especially ecology), economics, ethics, and a deep denial of readily apparent facts.
5. Too many people have bought into the fallacies of unlimited population and economic growth, expanding consumption/consumerism, and the ability of technology to overcome any problem whatsoever.

Fortunately, amidst all the short-sightedness and denial, many institutions and people are working hard to provide valuable education toward a sustainable, ecological path forward. Here are a few of them---

mahb.stanford.edu/

psmag.com/magazine/fallacy-of-endless-growth

williamrees.org/on-herman-dalys-economics/#more-261

thetyee.ca/Analysis/2020/04/06/The-Earth-Is-Telling-Us-We-Must-Rethink-Our-Growth-Society/

https://steadystate.org/

December, 2020

To ecologists, it's not a great mystery

Do an internet Search for this article:
"Protecting Nature is vital to era of pandemics", a wildlife Report in *The Guardian*.

A report issued by a scientific group which was convened by
the Intergovernmental Science-Policy Platform on Biodiversity and
Ecosystem Services includes the following conclusions (among others):
1. the increasing number and frequency of pandemics is due in large part to anthropocentric damage to natural ecosystems;
2. in addition to *reaction* to pandemics, **prevention is crucial**; and,
3. leading experts in ecology, wildlife biology, epidemiology, and similar fields have made the above points clear, but there has been little government action.

If you'd like to see the full report, do a Search for this:
IPBES Workshop on Biodiversity and Pandemics Report_0.pdf
The Executive Summary starts on page 2, and in it, needed actions are capsulized.

Politicians have been tinkering around the edges of the pandemic problem for decades (long before Covid). Emphasis almost always is on a **reaction** to an outbreak. Not enough effort is put into **prevention**. <u>That involves significantly reducing our negative impact on the natural environment</u>. We have to stop pretending we're doing that already. Not only politicians have to stop; all of us must.

September, 2021

Reframing the Crisis

While anthropogenic climate disruption is a massive problem requiring immediate attention, let's not marginalize other assaults on the ecosphere. Loss of biodiversity and other related disasters are interconnected to climate change, and every bit as important.

At the root, the ongoing crisis we're facing is as much a question of ethics as anything else. The hubris of humans in relation to other species and the environment is (in general) beyond the pale. We have to correct that if we expect to survive and thrive in the future.

We do have a model or two with which to work. Indigenous societies around the world and certain Eastern religions have understood for thousands of years the importance of treating Nature as sacred. It turns out they are on the mark.

As Barry Commoner said in 1971, "Nature knows best.". Our technology has accomplished amazing things, but sorry to say, it also has egregiously damaged our global habitat. As a species and overall, our attempts to improve human well-being have brought us to the beginning of our own extinction. Unfortunately, that's true not only in the ecological sphere, but also in the social sphere.

It will take a paradigm shift in ethics if we are to survive and thrive.
Two books which point the way: *The Sacred Balance - Rediscovering Our Place in Nature*, by David Suzuki; and *The Web of Meaning - Integrating Science and Traditional Wisdom to Find Our Place in the Universe*, by Jeremy Lent.

Do an internet Search for:
mahb.stanford.edu/blog/the-nature-and-overshoot-crisis-reframing-the-discussion-of-climate-change-and-biodiversity-loss/

September, 2020

Should we continue with infinite economic growth after the transition to a Green economy?

Short answer: definitely NO. Even for a moment, I shudder to think of the indoctrination going on in the world today relative to how the theoretical Green Economy (G.Econ) will facilitate the continuation of perpetual economic GROWTH. Is the G.Econ a good thing for us? Yes, it is. Is continuing unlimited growth a good thing for us? Absolutely not.

When you have time, do an internet Search for "Is green growth possible, by Jason Hickel". That journal article tackles the nexus of the G.Econ and perpetual growth, and it does so with both qualitative and quantitative empirical evidence. It not only analyzes all the factors involved, but applies *synthesis* to them as well. Those factors include: Gross Domestic Product (GDP), Domestic Material Consumption (DMC), decoupling, re-coupling, material and energy throughput, green growth, theoretical Tech innovation, global material footprint, unlimited economic growth, bioenergy with carbon capture and storage (BECCS), the IPAT equation, Eco-Crisis, ecology, natural resources, political ecology, and more.

The title and first paragraph of this essay give a partial hint, but I won't reveal the full conclusion of the cited article here because it's worthwhile to read the piece. So, put on your thinking cap, buckle down, study it, and I believe you'll become a convert, so to speak. This subject might seem boring to the "average" person, but just the fact that you're reading this means you're above average relative to concern over the ongoing Eco-Crisis. So many solution options to consider – G.Econ, Circular Econ, Bioecon, Steady State Econ, Ecological Econ, and various combinations of those. If *organized* human existence is to survive and thrive, we have to get this right.

The piece by Jason Hickel (and Giorgos Kallis) is an "Open/Free Access" journal article, and is well researched. Given the years-long search for Sustainability (especially in the EU), it could be extremely helpful to policy makers in their quest for best sustainable options.

Unfortunately, the current belief of most Powers-That-Be appears to support the idea which claims we can have infinite growth provided we just "Go Green" and trust "The Market". I do see signs, though, which suggest such a belief is beginning to crumble. It needs to crumble; science does not support it.

May, 2021

A Most Interesting Question

The title of a journal article in <u>Ecological Economics</u>, Volume 185, to be published in July, 2021 is:
"Beyond ecosystem services and nature's contributions: Is it time to leave utilitarian environmentalism behind?".

From the "Highlights" of the article (preceding the Abstract):
"We plead for a paradigm shift, away from utilitarian, anthropocentric and dualistic conceptions of human-nature relations." (Fair use quote)

As *Homo sapiens* attempts to move toward sustainability and all things "Green", it seems to me that we're still missing an important underpinning to the success of those efforts. It's known as ecocentrism.

In the dozens of articles on the circular economy, the doughnut/donut economy, etc. which I've read, the ideology of anthropocentrism appears to prevail. It's a view which sees humanity both superior to and separate from Nature. That contradicts Reality.

In any case and in general, humans do seem to be filled with hubris... especially regarding nonhuman life. Such is a problem when we profess to be "going Green". In my view, there will be no successful sustainability without the adoption of ecoethics on a significant scale. Ecoethics will not be the prevailing philosophical standard without a paradigm shift in our view of the natural world. In the meantime, our efforts toward sustainability
- **though laudable and a good first step** - *still amount to tinkering around the edges of the socio-eco-econ-ethical Crisis.*

Nonhuman life has **intrinsic** value, not just utilitarian value. The Earth is not here for us to subdue. The spark of life in *all* beings is the spirit of Universal Consciousness.

As Barry Commoner stated in 1971 (in his book, *The Closing Circle*): "Everything is connected to everything else". We humans are not separate from Nature; we are Nature. We don't "come into the world" when we're born. We come from it. All beings are Universal Consciousness expressing Itself.

Without the above worldview, I don't see how we can live in harmony with Nature, or have any significant degree of sustainability. *Homo sapiens* did live in such a way for most of our 200,000 year (or so) history.

January, 2024

The Ecological Ignorance of U.S. Presidents

It borders on insanity that common folks in the relatively free countries which produce and own nuke weapons (**9 total*****) aren't persistently calling for a ban on such weapons. Even a number of high-ranking military folks, such as retired Colonel Andrew Bacevich, have stated nukes are unnecessary for the defense of any nation.
[*****Many more than 9 "host"** (or endorse) **U.S. nukes**]

The nuclear experts who publish the *Bulletin of Atomic Scientists* now have their "Doomsday Clock" set at 90 seconds to midnight.

All our recent Presidents are in favor of so-called "lower yield, tactical" nukes. Obama launched a 10-year program (continued by Trump and Biden) to expand and upgrade our nuke arsenal. <u>Such action demonstrates an ecological ignorance beyond comprehension</u>. Apparently, they think use of such weapons will have no serious or long-term deleterious effects on the life which survives such a war. According to the scientific fields of ecology, biology, climatology, and physics, such thinking is sheer nonsense.

 Decades ago, I'm fairly certain most people were up to speed on the highly probable effects of nuclear war and ***the aftermath of nuclear winter*** on the biosphere. Nowadays, we have several world leaders who have developed, or are developing, so-called *tactical* (rather than strategic) nuke weapons.

If green plants and oceanic algae - which are the base of <u>food chains</u> AND supply us with <u>oxygen</u> - can't get adequate sunlight for months (perhaps up to a year or two), then such will be an extinction level event for many animals, possibly including humans. In any case, famine would be widespread. If animals are subjected to an atmosphere overburdened with soot and other fine particulates from explosion updraft, populations will crash. Any animal with lungs would be severely impacted. Then, on top of everything above, there's the damage from radioactive fallout. Never mind the impact on fresh water supplies. Ionizing radiation is the ultimate pollution.

On top of all that, more production of nukes means more nuke waste (which produces ionizing radiation for thousands of years). The waste we have already is *temporarily* being stored at whatever site generates it. That's been going on since the mid-**1940's**. There have been big problems, and they continue. [The Hanford, Washington Nuclear Facility, where plutonium was manufactured for the WW II atom bombs, is still the largest hazardous waste cleanup site (Superfund Site) in the USA.]

Ike's "military-industrial complex" rolls on, and very few people seem to care, or maybe they feel there's nothing they can do about it. **Here's something you can do about it:**
https://www.icanw.org/join
And here's part of why you should,
https://www.icanw.org/nuclear_arsenals

Don't believe anyone who says nuclear weapons and nuclear waste are not a significant problem. Such statements are the height of ignorance, an egregious insult to civil society, and a stupidly dangerous risk to all life.

September, 2022

Eco-Footprint Calculators

https://www.omnicalculator.com/ecology

Several PhD candidates came up with this series of unique "calculators", meaning algorithms. The broad categories are: eco-footprint, renewable energy, and "other". So, for example, under "eco-footprint" one of the calculators is "flight carbon footprint". Another is "cryptocurrency footprint", and another is "tree benefit". You plug in some specific numbers, and they calculate the benefit or damage.

Under "other", here are only a few of the calculators: "Shannon diversity index", "fish mercury", and "tap water". In all the broad categories, these are easy to use and quite interesting.

Tools like the ones above are helpful in revealing both our positive and negative impacts on the ecosphere. Sometimes such impacts simply are not readily apparent. I hope you will spread these far and wide. Thanks.

April, 2022

With several exceptions, there's not much to celebrate on Earth Day 2022

I remember well Earth Day 1970. That year I was teaching Earth Science in Klamath Falls, Oregon. People all around the USA (and probably elsewhere) were joining the Environmental Movement. The EPA had been created. As well as a number of other environmental laws, an updated Clean Air Act was passed in 1970. In many regards, it was a time of optimism. It seemed to be a paradigm shift for the better.

Except for a small number of countries/areas around the world (the USA is NOT in the exceptions), fifty-two years later we have the following:
1) in general, GDPs continue to expand, and that correlates directly to increasing ecological damage;
2) men like Putin apparently possess almost zero knowledge of ecological principles (never mind ethics);
3) like most other countries, the USA has no Fed Dept. of Sustainability;
4) Biden is re-opening drilling for fossil fuels on *public land*;
5) cleanup on the most toxic sites in the USA (Superfund sites) practically is at a standstill;
6) the petrochemical industry is pumping out more single-use plastic items than ever before;
7) the climate situation continues to deteriorate;
8) instead of being proactive and thus addressing the ***ecological causes*** of pandemics, governments continue to be reactive when fighting disease;
9) the Powers-That-Be are not adequately addressing biodiversity loss and numerous other eco-problems, most likely because they don't understand such problems are an *existential threat* to humans; and
10) there still is no permanent solution to the problem of millions and millions of gallons (or in many cases, pounds of solid material) of nuclear waste, which now is being temporarily stored in every State in the USA. [Some of it is escaping into the soil, the water, and the atmosphere. Such waste will remain radioactive for thousands of years.]

Given all of that, one wonders if perhaps we should change the scientific name of humans (*Homo sapiens*) to *Homo moronicus*. Just kidding, sorta.

To all the exceptions I mentioned, keep going with your good work. It must be discouraging at times, but you can prevail. Sustainability still has a chance. KUDOS, and thank you for your efforts.

Economics

November, 2013

New World Order...Again

Below is a comment I left on CBSNews.com, in response to a brief comment in the Evening News by John Miller about the New World Order (NWO). The recent LAX shooter had yelled out something about the NWO, and analyst Miller stated something to the effect that "conspiracy theorists" believed in the same thing, and in "black helicopters" and the like. That's the background for this response of mine---
[For this article, I've expanded the original comment a bit; additions are in brackets.]

……………………………..…

For Mr. Miller---RE: the New World Order---

1. You may or may not know, but former President Bush (Sr.) told us that we could look forward to a "New World Order". It's not a conspiracy theory.

2. Forget the "black helicopters". I suspect that very few believe in that scenario. The one of which Mr. Bush spoke is much worse...and again, it's NOT a conspiracy. It's right out in the open.

3. The actual New World Order involves the creation (ongoing as we speak) of a world economic structure run by transnational, mega corporations (including mega banks). [The results already are evident: neocolonialism in the Southern Hemisphere; manufacturing and jobs beginning to disappear in the Northern Hemisphere; Financialization (nothing but paper) replacing productive work; competition shrinking; and an exponential transfer of wealth from the Middle Class to the 1% Upper Crust. That's the real NWO.] Again, it's out in the open for all to see.

4. As David Rockefeller stated publicly in 1991 (paraphrased due to memory): "Surely it is preferable to have the world run by [business] elites than by nations and the corrupt politicians of the past."...that's the goal of the Corporatocracy...and it's not a conspiracy. The Oligarchy believes it's simply

good business...at least, that's my opinion of what they believe.

5. The Corporate Media love to conflate these two different New World Orders, and call the whole mess a "conspiracy theory". That phrase is used by Oligarchs to squash serious discussion of controversial subjects. But in the last several years or so, we commoners (I believe) have seen through that ploy. he propaganda won't work anymore. More and more, people are realizing that NAFTA, CAFTA, the WTO, the IMF, the World Bank, [Credit Default Swaps, Mortgage-Backed Securities, Financialization,] etc. are bad for not only the USA, but the entire world as well.

6. Thus, my suggestion is for the Media to stop using that tired phrase ["conspiracy theory"]. Your credibility is at stake. [That's already in the toilet, but perhaps if you stop, then maybe you'll get back some credibility.] In many cases, we just don't believe you anymore.

May, 2016

Why many people are pissed off

The next time you're on YouTube, do a YT Search for "What is the Precariat Guy Standing TEDxPrague". This brief, condensed talk (only 12.5 minutes) really does explain a lot, and not just in the case of Europe.

Here's why many people all over the developed world are pissed off---they've been forced into the Precariat socio-economic class by Globalization, Financialization, and the Plutocracy. The Precariat class lives with severe uncertainty in terms of job security, income, and quality of life. Now, years after this talk (in the YouTube clip), things are much worse. We all need to stop drinking the Edward Bernays style Propaganda Kool Aid coming from many Politicians, Plutocrats, and Oligarchs. Too many people are either in poverty or on the brink of it. They're employed, but the wages and benefits are not sufficient for a decent life.

Professor Guy Standing holds a PhD in Economics from the University of Cambridge (founded in the year 1209...and 1209 is not a typo:). He has worked for years developing the concept of the "Precariat", and is co-founder of the Basic Income Earth Network (BIEN).

November, 2014

Lori Wallach is a Genuine American Hero for Exposing Globalization

She works for Trade Watch at Public Citizen.org, and has a complete grasp of the abomination known as the Trans-Pacific Partnership (TPP), a so-called "Free Trade Agreement". Go to the link below to see why this Agreement never should be approved by our country.
http://www.citizen.org/TPP

About 500 to 600 attorneys from mega corporations essentially have created the 29 or so Agreements that form the core of the TPP. Only five of those Agreements/Chapters have anything to do with traditional Trade. The rest will benefit the Super-Rich at the expense of common people. The only reason we know this is because a whistleblower leaked the documents not long ago. Prior to that, it was the most secret Agreement in the Trade arena. Not even the U.S. Congress had access to it.

Very shortly, **Obama and McConnell will be working together to get the TPP approved here as quickly as possible.** They've said so publicly. This Agreement, if approved, <u>will offshore millions more jobs from the USA</u>, threaten the safety of our food, further threaten the environment, boost the cost of medicines, and protect mega bankers from prosecution to an even larger degree than presently. That's only a few of the detriments we'll experience as a result of this bonanza for mega transnational corporations.

The TPP also will allow ANY corporation that believes its profits have been diminished by any U.S. law to sue the Gov't in a *quasi-private* international tribunal in order to get compensation. In other words, corporations will be able to raid the U.S. Treasury if, for example, some safety law forces them to spend any money in order to comply. [Any judgments can be enforced by Trade sanctions against the USA.] **Obama and McConnell** certainly won't tell you, but that's one of the things they're working to get approved in our country.

Our puppet politicians and the Corporate (Mainstream) Media for years have propagandized us into accepting the falsehood that there are no alternatives to Corporate "Globalization". Lori Wallach and others have shown that is a bald-faced lie. For dozens of informative Video Clips on the subject, do a YouTube Search for her name. She has been exposing the whole phony business for **twenty years**. Kudos to you, Lori Wallach. We all owe you a great debt. With the aid of the work you've done, we peacefully can dethrone the Oligarchs.

STEP ONE: stop drinking the Propaganda Kool Aid coming from mega corporations and the politicians in their back pockets.

p.s. The upcoming TAFTA, the Trans-Atlantic Free Trade Agreement, is even worse than the TPP. [NOTE: TAFTA died on the vine (so to speak) due in part to the efforts of Ms. Wallach and others like her.]

August, 2019

BIG News on the Economy

On Wednesday night, 8-14-19, in their News broadcasts, the Corporate Media finally admitted that a Recession "may be" in the near future. Isn't it amazing how we (according to them) went from "our booming economy" to a Bust "may be" right around the corner?

Of course, the Corporate Media have known that for at least a couple of years. Their News isn't "fake", but it's often *incomplete*. The metrics for the next Crash have been in place for, at a minimum, the last three years. News of a "booming economy" meant a "boom" only for a relatively small percentage of income earners. The underlying, base economy has been pretty much a disaster. Half the workers in this country are either in poverty, *or bordering on it*. HALF.
We're talking about workers, not people on assistance/welfare.

The greatest transfer of wealth (to the Upper Crust) in the history of humanity is still in full swing. The so-called "recovery" doesn't apply to the majority of Americans.

Here's all you need to know for the near-term future: there's no "may be" or maybe (see the first paragraph above) about the coming Crash. It's about 99.9% guaranteed. We're overdue for it. Again, if you don't believe that, simply look at the time intervals *between* Booms and Busts from the 1973 Recession to present day. It won't be only here in the USA; it will be worldwide. It won't be a minor downturn. It will be a major Crash. All you can do is plan and act accordingly. [NOTE: several months after this essay first was written, the "Black Swan" of Covid-19 engulfed the world.]

What does "plan and act accordingly" mean? Following are my main suggestions for weathering a severe economic downturn.

1. Pay down as much of your debt as possible. When the Crash hits, your income and purchasing power may go down, but debt payments likely won't.

2. Stock up on *extra* everyday household goods...not only food, but paper goods, OTC meds, toiletries, anything you use regularly. Include fresh water (jugs/bottled) on that list. Credit will dry up for a time (maybe a week, or two, or ?), and businesses run on credit, not cash. That means delivery of goods to stores will slow way down or stop... temporarily.

3. Keep some ready cash handy. How much depends on your comfort zone. Banks very well may limit withdrawals for a time.

4. Hang on to (or obtain with cash) hard assets. Their value will increase when inflation hits.

5. Don't take on more major debt, *unless it's for an asset that you can sell relatively quickly.* When announcing their "news" about the Recession, the Media suggested that you perhaps (because of current lower interest rates) should now buy a house or a car. That's wise advice **only if you can pay cash and have no extra debt**. Corporate interests want you to take on more debt...don't.

6. Have a small food garden in your backyard, or do container gardening on your patio, porch, balcony, or even in a sunny room in your house or apartment.

No politician or political Party in the USA has eliminated the Capitalist Boom and Bust Economic Cycle. So, plan ahead and be prepared.

October, 2022

This might be the beginning of the end of Corporate Globalization

More and more, people with influence are recognizing the destructiveness and failure of neoliberal globalization. It's about time. **Do a YouTube Search for: The Failed Age of Globalization and a New Era of "Homecoming", Amanpour and Company.** The solution offered hearkens back to E.F. Schumacher's book, "Small is Beautiful"; at least, it's akin to that. Again, IT'S ABOUT TIME. This could be the beginning of the peaceful overthrow of global corporatism. Let's hope so. Perhaps policy makers will start seeing that all those "cheap" goods have <u>social, economic, and environmental</u> costs which have not been taken into account.

Incidentally, the goods coming from Asia actually are not cheap in price. Why? Because the quality of most of them is so poor that they have to be replaced about ten times more often than those made with durable materials, quality assurance, and quality control. Plus, as has been shown over and over lately, with great distances come monstrous supply chain problems. Suddenly the price of an item jumps up because it's scarce.

Finally, corporate globalization must go because mega companies (in general) a) don't really care much about customer satisfaction; and b) often evade accountability when they farm out manufacturing overseas. International Trade can be responsibly done, but the WTO (World Trade Organization) and the so-called "Free Trade Agreements" are an utter disaster.

https://rethinktrade.org/toolkit/hyperglobalization-and-supply-chain-failures/

Small is beautiful. Let's have an economics as if people really mattered to the Powers-That-Be.

December, 2023

OIL, Plus the Target Date for the Global Financial [and thus Economic] RESET: 2030, Plus Necessary Minerals

A few years ago, the IMF launched a ten-year Plan to reshuffle the world's reserve currency system. It became known popularly as The Global Currency Reset. For options other than the IMF version, do an internet Search for: globalresidenceindex.com/the-global-currency-reset/ . [NOTE: don't confuse this Reset with the World Economic Forum's "Great Reset".]

Many, many factors are responsible for bringing about the need for a Reset, but here we are focusing on only one: OIL. When you can, do an internet Search for: "simon michaux copy of gtk reports".

The Establishment loves fossil energy because it has been relatively <u>cheap, abundant, and convenient</u>; plus, it has made possible the seemingly never-ending GROWTH of the economy. For about 150 years, we've had a glut of such energy. That's why we think such a situation is the norm. It's not.

The low-hanging fruit (so to speak) is almost gone. Most remaining fossil energy is locked up in shale, limestone, and tar sands. It's abundant, but not cheap or convenient to obtain.

We also need tons and tons of minerals (which includes metals), and they are no longer low-hanging fruit. We especially need them for the transition to Green energy, which currently is proceeding at a snail's pace.

Examine <u>all</u> of Dr. Michaux's MAPS (not just the first one) at the link above. They contain a wealth of information (especially for planners and policy-makers) relative to economies, currencies, wars, geopolitics, and more. By the way, Michaux has determined that the world contains approximately **1.4 billion** air-land-sea transport vehicles (according to him, a conservative estimate). Just for the Transport sector of our economy, it would take hundreds of years of new mining to obtain necessary minerals for our Green transition. The current plan for "Going Green" needs to be adjusted.

Yes, we can recycle most minerals from batteries, solar panels, wind turbines and the like; however, nothing can be recycled from items *which are not yet built*. New mining will be required, and lots of it. It takes about sixteen years from the idea of a new mine to opening day of operation. Sixteen years.

Nothing previously written means we should abandon our necessary transition to a Green economy; but it does mean the current approach needs to be modified. Michaux has many, many suggestions as to how. Currently, he works for the Geological Survey of Finland, and has said in an interview many of the policy makers in Finland are intently listening to him.

January, 2024

MAGA should become MAMA: $50 TRILLION has been taken from Main St.

When convenient, do an internet Search for this essay: "Make America Mad Again, Common Dreams".

Mad about what?...trickle-down economics, that's what.

For a bit over forty years, the USA has encouraged/supported "trickle-down" economics. It has been a resounding success...for the Super-Rich. They took the gold from the mine; and Main Street got the shaft.*** There's nothing "Great" about such economics (but Trump thinks there is).

***NOTE: "the shaft" took the form of essentially stagnant wages, reduced employment benefits, job insecurity, often junk consumer goods, and jobs being shipped overseas or south of our border. For most workers, the "American Dream" became the American nightmare. The ultimate insult was that most Republicans AND Democrats supported it all.

At the link described above, a *Time* magazine article which explains all this is discussed/analyzed by Paul Buchheit, an economic justice author. He does a great job in this opinion piece of showing how MAGA ignores the reality of the last forty years.

To put $50 trillion into perspective, let's compare the "number" one million to the "number" one billion <u>in the framework of "time"</u>. One million seconds (of time) = <u>about twelve days</u>, and one billion seconds = about **thirty-one years.** As you know, the number *one trillion* = 1,000 times one billion.

On a related note---

Often the Upper 1% of income earners like to imply (or state outright) that they pulled themselves up the ladder of wealth by their own bootstraps (so to speak). Taking the vast High Tech Industry as an example, let's look at what really happened.

In the beginning (the 1950's), High Tech development was funded **almost entirely** by taxpayer dollars via the Defense Department. If it hadn't been for public money, which was massive for that industry, development would have been held back for who knows how long. [Even today, significant government contracts are awarded to High Tech companies and other mega businesses.] So, whose bootstraps?

Neoliberal economics (which basically is "trickle-down", mainstream economics) has transformed our economy into a tool for making the Rich richer at the expense of the common worker. _If wages had kept pace_ with the increasing *productivity* of workers in the USA, then today's "minimum wage" would be approximately $25 per hour. Instead, wages barely have kept up with inflation. For the most part, the average American worker is only treading water while the Super-Rich have enjoyed a 700% increase in income during the last forty years or so.

Income inequality is now over the top, beyond the pale.

December, 2022

How Most Billionaires Damage the Economy

When you have time, do a YouTube Search for the following clip: "Economists Debunk Top 5 Billionaire Myths, WIRED".

Unfortunately, too many people believe the myth that billionaires help the economy. This short clip (7.5 minutes) from *Wired* magazine explains why a "wealth tax" is necessary. The first *permanent* Income Tax in the USA was such a tax. It applied to only two percent of the population. Wages and salaries were not taxed, only passive income. Not long after that, for decades legislators were lobbied over and over by the Rich, and we wound up with the current tax set-up now in effect.

There's nothing wrong with being rich if one acts responsibly and helps society by paying their fair share. To get where they are, billionaires ***stood on the shoulders of society*** via education, infrastructure, networks, communication, agricultural surplus (food on the table), civility in our habitat and relations, law enforcement, courts, healthcare facilities, and fire departments. In addition, perhaps they had a bit of luck: maybe born on third base, maybe smarter than the average person, or taken under the wing of someone with wealth/influence/power. In any case, there's no such thing as a "self-made" billionaire. That's another myth.

Any super-rich person should be willing and happy to help out society by means of a wealth tax. We humans are a social species, and that includes those in the Upper Crust.

Science

February, 2024

Introduction, a Personal Note

As far back as I can remember, Science (especially Biology) and Nature have fascinated me more than anything else on Earth. Nothing else comes close. As young kids, my brother and I would catch polliwogs and tadpoles in the waters around Houston, Texas, place them in a make-do aquarium, and watch them grow into toads and frogs. Watching adult Leopard frogs catch insects with their lightning-fast tongue shooting out like a bullet, the fascination and awe continued. Perhaps the best sight was their incredible leaps/jumps in different circumstances.

We also caught and observed honeybees and the larger bumble bees, as well as all kinds of butterflies and moths. Then, too, there were the larvae, the caterpillars and eventually their cocoons. That was my early introduction to metamorphosis. Prior to their amazing change, caterpillars (before the cocoon stage) eat a phenomenal amount of vegetation. They grow so fast and so much that their skin has to be shed multiple times.

In later childhood, I patched together a homemade, glass ant farm and managed to stock it with a variety of the critters. I couldn't get enough of watching them build their home, manage food and water, and interact with each other. It all left quite an impression on a very young, budding naturalist. Just prior to teenage years, I became what's known as a "birder"; and it's still with me. Back in the early days of "birding", I'd roam fields and forests, observing the habits and interactions of those marvelous creatures.

Next came High School Biology class, and later, two different Universities with training/studies in agriculture, biology, geography, and ecology. All the above solidified my love of and fascination with the life sciences. Two Professors in particular (who are cited in the Dedication on page 2 of this book) at Arizona State University, Ira Judd and Virgil Baker, fired up my interest in ecology so much that it became a lifelong passion.

During extensive studies in ecology, an important point became clear to me. Science does not deal in "Absolutes" when it comes to *conclusions* in research studies. In the years since I realized that, it seems people on Main Street have developed a serious misconception regarding research in almost all fields of science.

When it comes to the **<u>conclusions</u>** of a study (<u>not the</u> **<u>measurements</u>** leading up to conclusions), Science and scientists usually deal in **degrees of probability**, not certainty. In other words, science is not Absolute. Why? There are too many unknown variables which cannot be accounted for in any scientific investigation. Because of that, when new discoveries are made in a particular area of research, the previous theory is likely to be modified or even discarded. With new research, past conclusions often change. **We should never forget**: for a long time, it was believed to be a "fact" that the Sun revolved around the Earth. After all, *anyone could SEE it happening* every day. "Proof" positive, it seemed at the time. Ooops.

People should not lose belief in Science just because it deals in degrees of probability (regarding conclusions of a study) rather than absolute certainty. Not even scientists are all-knowing. A number of theoretical physicists have stated (paraphrased) : humans barely have scratched only the <u>surface</u> of Reality.

2022, 2023, 2024

Reductionism, Holism, and Complexity Theory

I've stated what follows next in previous written works, but these concepts are so important they bear repeating. These ideas are critical to understanding why many ecologists believe those in Power are not correctly addressing our ongoing socio-eco-econ-ethical Crisis.

There's plenty of evidence indicating both public and private Powers-That-Be are, for the most part, relying *strictly* on the reductionist method of analysis when working on our current, worldwide problems. It appears little thought is given to the holistic approach. If such really is true, then efforts to mitigate or eliminate those problems most likely will fall short.

All the primary aspects of our Crisis are interconnected, and <u>THE main problem</u> is *ecological overshoot*. Ocean acidification, excessive natural resource depletion, pollution of water, air, and land, soil degradation, food insecurity, famine, decreasing fresh water availability, drought, climate change, social unrest, and more are all caused by humans engaging in eco-overshoot. This all means the problem is extremely **complex**, and that means addressing it must include a holistic synthesis. "Systems thinking" is required.

 Many people appear to evaluate concepts by means of *reductionism*. In that method of analysis, whatever is being examined is broken down into its constituent parts. Each part is thoroughly evaluated separate from the others, and then the various parts are added together to determine some sort of conclusion regarding the whole. Such a method can be extremely misleading when dealing with complex (not just complicated) systems. Complexity is *nonlinear.* [A motor vehicle is a linear system; it's "complicated", but the behavior of the "whole" is predictable by examining the individual parts. Thus, it's not a complex system.]

Instead, for complex phenomena (such as the Eco-Crisis, or even a single ecosystem, or a single living organism), the method of examination described in Complexity Theory, or its precursor, General Systems Theory, should be utilized. Why? The approach used in Systems Theory (the updated version doesn't include the word "General") and Complexity Theory is *holistic*, not reductionist. It uses synthesis rather than only analysis of parts. It is aware of **synergy:** a whole system <u>often</u> or arguably, always, is greater than the sum of its individual parts, which means the behavior of a whole system is usually *unpredictable from the analysis of only its parts.*

A bit of history may help to understand this whole idea. General Systems Theory has been been around for several decades, well over sixty years. It developed a number of concepts centered on "systems", as follows.

1) A system is a set of interconnected elements which interact with each other due to the relationships of those elements (e.g., a motor vehicle).

2) It has separability, meaning the system can be separated from its surrounding environment. It is <u>*distinct*</u> from its surroundings. Theorists in this field consider that to be an essential characteristic in the definition.

3) A system is either "open" or "closed". If it interacts with its environment (as in, for example, receiving energy or matter) in order to maintain its structure, then it's an open system. An automatic washing machine such as the one in most homes in the developed world is a closed system. It's distinct from its environment, and requires nothing from outside itself to *maintain its structure*. Any system containing life (even a single organism) is an open system. Nonliving open systems also exist. An example follows: if a pot of water on a stovetop is continuously boiling, it's an open system. It's getting heat energy from outside itself in order to keep boiling. If a covered, non-boiling pot of water is just sitting there, it's a closed system.

4) The setting, the circumstances, the backdrop, the conditions in which a system exists are all crucial to understanding complex systems. In other words, the **context** has to be grasped to properly analyze a complex system.

Especially in physics and chemistry, General Systems Theory (GST) started off with a bang, but after several years it became apparent that a serious problem existed regarding separability and ecosystems, as well as simpler biological systems such as a single organism. [Given that the acknowledged Father of GST is considered to be Ludwig von Bertalanffy, a theoretical *biologist*, the situation is somewhat ironic.]

Especially regarding ecosystems, the problem was difficult to overcome. When doing analysis (or synthesis), GST theorists were having trouble separating such systems from their surrounding environment. That's because a natural ecosystem *IS* the environment. When analyzing human impact on an ecosystem, contextual behavior is everything. Plus, emergence of unpredictable behaviors and properties is common. Such emergent properties of a complex system can occur when individual elements and/or subsystems interact with the whole. Emergence easily can be overlooked. In particular, large ecosystems are the most complex systems on Earth, and that includes the *production* of things made in factories by humans.

Partly because of what's described above, in the 1980's and early 1990's several researchers developed Complexity Theory. In my opinion, it has an even more holistic approach to analysis than does GST. Chief among the new theorists was Prof. John H. Holland, who came up with the idea of "complex adaptive systems" (cas). Google him and "cas" when you have time.

So, why does any of this matter? As stated in the first couple of paragraphs in this essay, many ecologists (including myself) believe most of the Powers-That-Be are on the wrong path in their approach to addressing our ongoing Crisis. They don't seem to see the interconnections of all the various symptoms of ecological overshoot. In fact, they act as though climate disruption is the only thing we need to worry about, and that everything will be fine by 2050 if we continue to transition to Green energy. <u>Discussion of reducing economic throughput is not even on the table</u>. Such an approach will result in devastating future suffering. It's a path to ecocide, and it must be abandoned if organized human existence is to survive and thrive.

Ethics

December, 2010

The Best Thing in 2010

[2024 NOTE--- In 2010, Wikileaks published thousands of "Iraq War Logs" which exposed hundreds of U.S. Government misdeeds. All of those misdeeds could be considered unethical; some were illegal as well.]

The best thing, that is, relative to peacefully fighting the Corporatocracy, or the Military-Industrial Complex, or the Oligarchy, or the Plutocracy... or whatever one wishes to call it. That "best thing" was the action by Wikileaks.

Many have disagreed already, and I suppose many more will. They think it was a "bad" thing. Their main argument seems to be: the leaks put lives at risk. The only specific life <u>at risk</u> that I've heard about so far is that of an Iranian person; but, there might be others as well. There are a few possible retorts to that argument, as follows.

1. It is not the job of Wikileaks to secure the safety of a spy. That job falls to the spy and the Government. Apparently they didn't do a good job.

2. To use the Government's own argument, sometimes in accomplishing a greater good, there is "collateral damage". Was it a greater good? Yes. We the People have a right to know what the Corporatocracy is up to, and what its adherents think of other governments. We have a right to know if they are up to no good, and too often they are. Cases proving it number in the hundreds or perhaps thousands. A few examples: the instances of damage to "downwinders" relative to above-ground atomic testing decades ago; the instances of damage to soldiers and others who were given LSD without their knowledge; the instances of damage resulting from Agent Orange during the Vietnam War; the instances of damage known as the Gulf War syndrome, almost certainly due to exposure to "depleted uranium" (that's government-speak for nuclear waste) and/or untested vaccines; and, the Tonkin Gulf incident in 1964, which did not happen the way the Government reported it.

3. How is this significantly different from when the CIA's Valerie Plame was "outed", probably by Dick Cheney via Scooter Libby? I don't recall too many folks (especially "Conservatives") crying foul over that action.

4. If we accept the commonly held belief that our Government must be able to keep secrets from the People in order to operate effectively, then where do we draw the line...and how do we know that the agreed upon line is being properly observed by the government? Seriously.

5. Similar to #2 above: our central government, as well as mega corporations, have shown over and over that they are not to be trusted. We need to know as much as we can get our hands on in order to keep our health and freedom.

April, 2018

The End of American Hegemony is Now on the Horizon

It's difficult to predict whether that end still is a long way off or will happen somewhat quickly...too many variables. Nevertheless, it is finally visible. Despite what Oligarchs and those who are propagandized to the hilt may think, the approaching end of the USA's worldwide hegemony is a good thing for everyone except powermongers..

A few years ago the results of either a Gallup or Pew Research Poll (can't recall which) indicated that 80% of the world believed the greatest threat to peace was the USA. I suppose our Bully Boy, Reality Show President would call those results "fake news"; but then, everything he disagrees with is so named by him.

China, the rest of Asia (including India), Russia, Iran, large parts of South America (especially Brazil), South Africa, and now even Saudi Arabia, are so fed up with our Govt's HEGEMONY that they are waging an unprecedented economic/financial war on the USA. **They have a good chance of winning it.**

First came South America's rejection of our Government's longstanding policy of Neocolonialism relative to them. Next was BRICS (Brazil, Russia, India, China, and South Africa) and their financial revolt, and just recently, we have the PetroYuan, *essentially backed by gold.* Finally, many countries are allowing escape from the "almighty" Dollar via cryptocurrencies. Dollar-dumping is rife in many, many countries. [2024 NOTE--- The BRICS alliance now has many more members than only the original five.]

We live in interesting times--- we are witnessing the beginning of the end of American Hegemony... especially monetary/financial hegemony. Our Government and its Mega Bank Cronies will do ***anything legal or illegal*** to prevent that end. It all hearkens back to Manifest Destiny and the crimes committed under that "justification". The main difference between then and now is that our Empire is no longer land-based. It's now based on **control** via threats of force, both monetary/financial/economic force and military force. **Most of it is illegal, immoral, unethical, and unconstitutional.**

Starting a war won't prevent the end of our hegemony; it might only postpone it.

The Constitutional Republic of the USA is now in name only. *These days, we live in a Fascist State.* According to both FDR and Mussolini, Fascism is the Power of the State married to the Power of BIG Business. Undeniably, that's the U.S. Government both for quite a number of years and presently.

The days of genuine American Conservatism (a big part of which was non-interventionism abroad and non-speculative banking at home) AND genuine American Progressivism (a big part of which was a safety net for the most vulnerable and fairness for all) are mostly gone...long gone.

May, 2018

The Oligarchy Handbook

To my knowledge, no such Handbook exists; if it did, however, this would be a good outline or condensed version. NOTE--- Certainly not all of it, but much of the information below I've paraphrased from Noam Chomsky's superb book and documentary titled, *Requiem for the American Dream.* The editors of the documentary came up with this subtitle: "The Ten Principles of Concentration of Wealth and Power". I would call it "The Ten Steps for Instituting Corporatism in a Society". These concepts are evident not only in the USA today, but also in all industrialized countries of the world...some more than others.

1. Reduce democracy---
Reduce or eliminate the input and influence of commoners in the formation of public policy. The "little people" should be treated as "spectators" when it comes to government because they are incapable of comprehending complex matters. They should be distracted with consumer consumption and kept apathetic.

2. Shape ideology---
This begins with indoctrination in education. Critical thinking and any challenging of Authority should be suppressed. Except for basic skills such as reading and math, the minds of students should be molded to always accept Authority, wage slavery, war, the hierarchy of society, and a belief in the wisdom of opulence and luxury.

3. Redesign the economy---
Complete the move (started in the 1970's) to Neoliberalism, Financialization, and Globalization. In part, that means privatization of government infrastructure and services, and the promotion of worker insecurity. Separate worker productivity from wage increases. Keep wages stagnanOne way to do that is to promote the moving of manufacturing to emerging

economies with low wages and sweatshop conditions. Reduce or eliminate regulations applying to the Financial Sector so that income can be made without producing any goods or services...as with the use of bizarre financial derivatives and other financial instruments.

4. Shift the burden carried by Oligarchs---
Make the Rich and Powerful less responsible for funding government and for their own mistakes. Reduce their taxes, and promote "Too Big to Fail", "Too Big to Jail", "Bail-Outs", and "Bail-Ins". Increase government subsidies to mega corporations. Let the "little people" foot the bill.

5. Attack solidarity---
Reduce (through Edward Bernays style "molding of the public mind" Propaganda) the caring by individuals for the fate of others. Destroy unions. Clamp down on freedom of assembly, nonviolent protest, and any group that challenges the status quo.

6. Run the regulators---
Promote and encourage a revolving door between BIG Business and high-level government service, especially in the Financial Sector. Poo-poo the concept of "conflict of interest". Promise and provide high-paying, cushy, corporate jobs to regulators when they "retire" from goverment service. Strive for "Government-Sachs".

7. Engineer elections---
Continue the practice of having high-level political candidates selected, groomed, and financed by the Rich and Powerful. Continue to encourage gerrymandering and voter suppression/manipulation. In Primaries, continue to encourage the Corporate Media to slight or ignore any candidate not approved by the Oligarchy. Promote and encourage the discussion of personalities rather than issues. Continue to have debates run by a private corporation, thus making certain that no genuine Populist candidates are allowed to participate. Make sure corporations continue to be "persons", and thus are able to contribute to our political candidates (via "free speech" money) in a virtually unlimited fashion.

8. Keep the rabble in line---
Inundate the public with Edward Bernays style Propaganda Kool Aid all year, every year. Attack Organized Labor. Attack any Populist group. Promote nationalism rather than patriotism. Talk jobs, not profits. Never reveal that while the wages of the "little people" have remained relatively flat (after adjustment for price inflation) for 30+ years, our incomes have increased 700%.

9. Manufacture consent---
Mold the public mind to our benefit. Keep the Propaganda flowing--- Trickle-Down Economics really works, there's no conflict of interest in having a Mega Bank CEO run a Gov't financial department, Big Biz should have a Stay-Out-of-Jail card, the Boogie-Man is coming so we have to go to war, the type of Globalization we're promoting really is good for the public, the commoners can have either privacy or safety (but not both), etc. They've bought it all so far, and they will continue to buy it.

10. Marginalize the population---
Make society an amorphous mass. Attack and destroy groups critical of Power and Oligarchy. Discourage the linking of individuals into effective political action organizations. Denigrate them whenever possible. Encourage the belief that commoners cannot understand the complexity of the issues facing the world today. This all creates an unfocused anger which makes people more susceptible to our control. Gain control of the internet, which is the greatest democratizing force in existence today (and not good for us). Oppose net neutrality whenever possible.

If we follow these steps, we'll continue to be Masters of the Universe and in control of the little people.

January, 2014

The Failure of the Meritocracy: Twilight of the Elites

Chris Hayes has it exactly right. He's an Editor-at-Large for *The Nation,* (at least he was in 2012) and has a book out titled, *Twilight of the Elites: America after the Meritocracy.* What I refer to as the Corporatocracy (from John Perkins, *Confessions of an Economic Hit Man*) or the Oligarchy, Mr. Hayes calls the Meritocracy. It's the same kettle of rotting fish.

The word "meritocracy" was coined by a British social critic in the 1950s. It was used in a satirical way to describe how the elites rig the game and make rules in their favor. Basically, the whole thing boils down to this:
1. in the ethereal atmosphere of the elites' world, the most valuable characteristic of Humankind is high intelligence;
2. the most valuable use of that intelligence is deemed to be the making of vast sums of money;
3. other considerations, such as a social conscience, egalitarianism, democracy, compassion, and the like are deemed to be unimportant...or even hindrances.

Though the Brit who coined the term meant it in the vein of *1984,* or *Animal Farm*, or *Brave New World* (in other words, dystopian societies), elites have (supposedly by mistake) adopted the concept as a model for society. As a result, they have come to view themselves---because most of them are super-intelligent---as a class apart, superior to the Average Jane or Joe in every way. They also have developed a strong sense of solidarity within their Class because they are, in their minds, vastly superior to the "little people". [I was stunned when the International Head of BP, in a national TV news interview, referred to the Gulf residents as the "small people". The man reeked of condescension...and I think, knew it.]

This all has resulted in a situation in which the elites, who are great Institutionalists, wind up protecting the people in the institution (public or private) rather than the people whom the institution serves...the overwhelming majority of any population. Those of us in the lower socio-economic classes are viewed as Insurrectionists (of a sort), and there is little or no concern for our well-being. Example: the cover-up (or ignoring) of the Sandusky affair at Penn State by the hierarchy of that institution. Example: the opposition to Cesar Chavez when he protested the lack of protection for farmworkers from pesticides, back when pesticides were barely regulated.

Those at the bottom of the socio-economic ladder have experienced the negative effects of the Meritocracy for *decades and decades*. What's different now is that the Upper Middle Class, since about the late 1990's or 2000, also has been experiencing those negative effects...especially economically, but also socially. Wages/salaries have largely stagnated, people have to work longer in order to survive retirement in a somewhat comfortable fashion, economic "bubbles" often have negatively impacted retirement funds, the National Security State is becoming more and more demanding and intrusive, wealth is being transferred from the Middle Class to the Upper One Percent at a truly astounding rate, and on and on.

The Sea Change in all this, the thing that has the elites very concerned, is the fact that the Upper Middle Class (and even many of those at the bottom of the ladder) is demonstrating that it will not put up with business-as-usual. Sometimes actively, sometimes passively, they are expressing their extreme displeasure. Faith in institutions of all kinds is in a free-fall decline, and most of those institutions are controlled by elites.

The Meritocracy/Oligarchy knows that something bad for them definitely is in the wind. Because they're mostly short-term-gain thinkers, I predict that in the not-so-distant future we'll have more economic bubbles and more "too big to fail" bailouts, or something even weirder, enabling the Upper Crust to scam even more billions. Financial "derivatives" may be instrumental in that scenario, again. I hope I'm wrong.

May, 2023

In the USA, Class War is Accelerating

For over forty years, the Establishment has been favoring the upper 10% of income earners with tax cuts, tax loopholes which go unplugged, subsidies for big corporations, government contracts, lack of prosecution for wrongdoing, and more. For the last few-to-several years, Main Street has become more and more aware of all that, and is finally pushing back with demonstrations, strikes, boycotts, and the like. They have done so despite the fact that the Establishment often criminalizes those who peacefully object to being ripped off or exploited. On top of all that...

The Super-Rich CEOs and corporations have realized what's happening to the American Empire - it's crumbling. As a result of their realization, they now are making extra efforts to garner more wealth in order to further secure their future in the Post-American-Era. Those extra efforts include <u>increasing profits even more at the expense of workers</u>, buying influence with politicians more than usual (they spent one billion dollars on the last mid-term elections), treating customers much like livestock (denying claims re faulty goods or services, ignoring safety concerns, etc.), buying back their own stock so shareholders get that money instead of workers, and in general, cutting corners to the max at the expense of Main St.

Then, too, there's this...
https://www.democracynow.org/2023/5/24/
jeffrey_sachs_defense_spending_us_debt
In the current "Debt Ceiling" talks, *neither Biden nor McCarthy* is mentioning or considering cuts to Defense spending. The USA's spending on defense-related items constitutes **40%** <u>of the entire world's spending</u> on such items. Our Defense budget is larger than that of *the next ten nations combined*. Most politicians in both major Parties support the concept of perpetual war. So, the Defense budget is sacrosanct. Plus, increasing taxes on the super-wealthy is not on the table.

Instead, the Repubs want to cut spending on items which benefit the poorest Americans. Is that because they believe the Poor are undeserving of help? If not that, what is it? It can't be concern over fraud in some welfare instances. Why? Because they don't seem concerned at all about rampant mega corporation fraud.

All of the above has accelerated the ongoing Class War. The bought-off politicians on both sides of the aisle largely are to blame. So are many/most of the mega corporations and their Super-Rich CEOs. If you agree, think hard before casting your vote in the next national election, and before buying another piece of manufactured rubbish made in China.

And so it goes, all during the ongoing, socio-eco-econ-ethical Crisis.

January, 2024

Quote: "...whether God-given, biologically endowed, or just taken for granted."

This masterfully written essay (at the link below) clearly shows aggregate humanity is on a most destructive path re nonhuman animals. [Search online for: great transition gti forum/solidarity animals crist] The author inspires the reader to the Nth degree. At least read the first three paragraphs. **Judith Crist is a wordsmith of the highest order.** Reading her article is well worth the time.

Crist goes on to unravel/unpack the paradox of our love of animals existing simultaneously with the structural violence we impose upon them. Then, too, she brilliantly recounts the history and heritage of John Rodman's "Differential **Imperative**" - how we, in order to soothe our conscience, assign inordinate value to the difference between us and other animals. [I say "other animals" because we're part of the biological Animal Kingdom; obviously, we aren't plants or microorganisms:]

Next, Crist relays how indigenous societies have (in general) treated nonhuman animals with respect, compassion, and the dignity they deserve *while using them for human purposes*. It's a lesson for us all.

Finally, she explains how mistreatment and misuse of animals has spilled over to include human-to-human interaction, and our degradation of the ecosphere. Someday we'll all learn: "everything is connected to everything else" ~ Barry Commoner, 1971 (another masterful wordsmith).

March, 2014

Signs indicating the Elites in our Fed Government have lost it

As reported on *Democracy Now!* ---

1. The FBI has deemed Ryan Shapiro's PhD Dissertation on Animal Rights to be a *threat to national security.* Ryan, an Animal Rights activist, is a PhD candidate at M.I.T. He also has submitted more Freedom of Information Act (FOIA) requests to the Feds than any other American. Who knows, perhaps the tagging of his Dissertation was Government payback for all those FOIA requests. I don't know how likely that is, but it's certainly possible.

2. In 2006, Dubya Bush signed into law the Animal Enterprise Terrorism Act. As a result, animal rights activists who go undercover into Big Ag factory farms and film deplorable conditions and gross mistreatment of livestock may be charged *with terrorism.* Nonviolent, peaceful activists who stage sit-ins or lie-downs at such factory farms are not just practicing civil disobedience protests anymore; they may be charged with terrorism as well.

3. The Feds have tagged the underground Animal Rights activists (not the out-in-the-open protesters) and the underground environmental protesters as the *number one domestic terrorist threat.* So, if you break into, say, a fur farm at night and release all the critters, you're part of the number one terrorist threat in America. Really? Imagine if you stole an apple in the process. Admittedly, a few of these eco-activists and animal rights activists who are clandestine have burned structures to the ground or otherwise damaged the property of Big Business in the past...but, the number one domestic terrorist threat? I don't think so. As far as I know, not one person has been injured as a result of their illegal actions.

4. In the course of monitoring the peaceful "Occupy Houston" protest not too long ago, the FBI discovered a sniper plot against the leaders of that rally. The FBI kept that fact from everyone.

5. In a memo and a scenario similar to a portion of the old COINTELPRO, the FBI describes how peaceful Animal Rights activists can be neutralized by spreading false rumors that certain ones of them are FBI undercover agents. More lies from the Fed Government.

6. Ryan Shapiro (see number 1. above) is suing the NSA, FBI, and DIA (Defense Intelligence Agency) for their refusal to disclose the USA role in the imprisonment of Nelson Mandela. Shapiro also wants to know why the Government kept Mandela *on the Terrorist Watch List until 2008*, which is another indication of how the management of all this unwieldly "terrorism" info is really beyond the capability of the sprawling Fed Government. The right hand often doesn't know what the left hand is doing. I think there's a misconception among the public that the central government is one big monolith. In reality, it's thousands of semi-autonomous institutions which often are competing for turf and funding. The result is what you see today, some of it good, much of it miserably bad.

February, 2024

Why the factory farming of animals (and animal lab experimentation) is an obscenity

https://tamerlaine.org/factory-farming-in-pictures

1. Family farms, shrinking in number, still outnumber factory or "industrial" farms by far. BUT, the total number of animals on factory farms is greater than the total on family farms. A ***single factory dairy farm*** *can have anywhere from* **700 to 25,000** milk cows...on ONE "farm". Dairy cattle! These cows are never on pasture. They live in giant warehouses and/or crowded feedlots. There are about 9.4 million milk cows in the USA; the majority of them are never grazing on pasture their entire lives. [NOTE: in China & a few other countries, there are factory dairy farms with up to **100,000** milk cows on each one.]

2. Hogs on factory farms live their entire lives indoors, most of them in steel bar <u>crates</u>, not roaming around in a big indoor arena. Sows live in gestating crates barely longer and taller than the animal...<u>their entire lives</u> (which usually is only a few years). [See the photo at the link above. Not counting dogs, hogs are the most intelligent of all farm animals. In general, they can do simple math better than a 3-yr-old human. Like all livestock, they are social creatures, & have emotions.]

3. Poultry on factory farms are never outdoors anymore. They're raised in crowded crates in giant sheds, several packed into each crate, and crates stacked upon each other.

4. Lab animals--- I can't even go there--- many of the experiments are cruel beyond belief, done in universities with full Government funding. Mice, cats, dogs, primates, and a few other mammals are subjected to inordinate suffering. <u>Sentient beings treated like nonliving objects</u>. There are a few corporations which do nothing but raise "crops" of these creatures, thousands of future victims for sale. It seems many humans pretend all

animals except us are too dumb to be aware of their own suffering. I guess such thinking soothes their consciences.

In most of these factory operations, cleaning of animal wastes is done automatically, which means it's incomplete. Because of the massive numbers of animals, infected/sick ones often get overlooked and die without medical treatment.

The philosopher, Kant, once wrote (paraphrased): we can judge the character of people by their treatment of animals.

Factory farming (especially of animals) is an obscenity. It's a gross insult to civil society, sentient beings, and Nature. ***We are capable of far better behavior.*** Studies have shown that "small farms can feed the world". Google it. [You'll find arguments and studies supporting both sides of the question. The disagreement largely revolves around the definition of the term, "small farm".]

February, 2024

An apparently much-needed BIOLOGY lesson for the Alabama Supreme Court

When a female human egg is fertilized by a male human sperm cell, the result is known as a *zygote*. [The same is true with all other members of the biological Animal Kingdom.] A zygote still is a single cell. It has no brain, central nervous system, eyes, heart, etc. [Yes, that should be obvious and known by everyone. Given the Court's decision, however, thought I'd better make sure:]

When the fertilized egg divides into two identical cells, and those two divide, and on & on, the result is known as an *embryo*. While an embryo, the group of cells appears no differently than does a group of cells from any number of other animals at the same stage of development. It's blob of **undifferentiated cells**. It's not a child, not even close. There's no brain, no developed systems found in a fetus, no sentience, no consciousness, nothing but a group of identical cells. Yes, an embryo has the potential of becoming a human life; *but it's not yet such.*

Approximately eleven weeks after the last menstrual period in the mother, an embryo begins the phase of *cell differentiation*. Body systems develop with specific organs. The eventual result is known as a *fetus*. The differences between an embryo and a fetus are monumental and relevant to ethics.

When an embryo is frozen (which happens in the IVF*** process weeks prior to the beginning of the change to a fetus) cell differentiation doesn't begin until after unfreezing.
[***IVF = in vitro fertilization]

To treat/view an embryo as a child is absurd. To make things worse, one of the justifications for the decision was a citation from the Christian *Holy Bible*. I was unaware of Alabama being a theocracy, or at least, acting like one. There's no way that's appropriate...or even legal I would guess.

By the way, when you do a Google Search for questions regarding an "embryo", about one-third or more of the answers deal with a FETUS, not an embryo. So it seems some Tech Whiz "Kids", the ones who write algorithms for this Search topic, *also need this Biology lesson*.

End of lesson. *;)*

War

September, 2014

Current Wars of Obama's Fed Government

The USA's Fed Government is involved in the following wars. They are wars because Obama is having the U.S. Military (not Law Enforcement personnel) shoot at people and kill them. That's war. These all belong to Obama now.

1. The Third Iraq War, the one launched just recently.

2. The ongoing War in Pakistan, the one where the highest levels of our Government order Hellfire missiles to be shot (from drones) at SUSPECTS.

3. Ditto the War in Yemen.

4. Ditto the War in Somalia.

5. The War in Afghanistan, our longest war, and the one nobody is quite sure anymore why it is that we're still there. Obviously, the Carlyle Group, Halliburton, Raytheon, CACI, GE, Boeing, McDonnell-Douglas, and several other transnational mega corporations are enjoying tremendous benefits from that never-ending war, but surely that's not a good enough reason to STILL be there. Is it? Well, yes, if you are a member of the Plutocracy in a soft Fascist State such as the USA---where the public constantly is being bombarded with Edward Bernays style propaganda---then it is a good enough reason. Follow the money; *war is Big Business*...very big.

Those are the wars of which I am aware and/or can remember at the moment. Perhaps there are others as well. Some may attempt to argue that those five wars really are *all one war--- the War on Terrorism.* It's a tempting argument, and we certainly have been propagandized enough to believe it. If, however, we engage in some critical thinking, that argument quickly falls apart. Consider the following.
1. The fanatics we're fighting are all in separate entities; there is no "central command". There is no overall group with the title, "Terrorism". So, for example, the Taliban (whose leaders we hosted here in 1997 in order to

negotiate a pipeline in Afghanistan) have completely different objectives as compared to ISIS. They only want to run Afghanistan, as they were doing in 1997 when we invited them here for a business deal. [Query: did our Government consider them nasty brutes then? If so, why were we treating them as potential business partners?]

2. The propaganda is this (especially among so-called "Conservatives"): there is one group we're fighting; it's "the Muslims", in particular, the fanatical element of the Muslims. Really? So, are they Sunnis or Shiites? The fanatics, the extremists in Islam (not the mainstream), are at each other's throats. They are not one group. They appear to be worried more about each other than about us. Plus, from what I can see, the Taliban don't give a damn about anything except Afghanistan. Us still being in that country is complete nonsense. To think of "the Muslims" as a monolithic, homogeneous entity also is nonsense.

3. Specifically, whose military are we fighting in this supposed "War on Terrorism"? The people we're fighting are not in an Army; they are a bunch of heterogeneous criminal gangs (of a sort). They each have their own agenda. Should any of them ever attack the USA again, *it is a matter for Law Enforcement, not our Military.* Most all of the plots against our country have been thwarted by LAW ENFORCEMENT Agencies, not our Armed Forces. Soldiers are not cops, and shouldn't be expected to fulfill that function.

Finally, the five wars listed above are all illegal, unconstitutional wars. Three uses of the military are permitted by the Constitution: to repel invasions (of this country); to quell insurrections (in this country); and to enforce the laws of the land (in this country) when minor rebellions make that necessary - for example, the Whisky Rebellion of the late 1700s, or the refusal to integrate schools in the South in the mid-sixties. No law trumps the Constitution, so the War Powers Act cannot be used legally to justify going to war for some reason other than the three scenarios listed above.

That leaves only TREATIES to justify wars outside the USA (if we haven't been attacked at home). ***Treaties do trump the Constitution.*** IF that's what's being used to justify the five wars above, I'd like to see those Treaties...especially the one with Somalia or any of its neighbors. [As to

Afghanistan, it wasn't the Taliban who attacked us on 9-11-01. Their refusal to turn over Osama supposedly justified us going to war with Afghanistan. The truth is: the Taliban replied to our Fed Government's request by saying essentially this: show us some proof of his guilt and we'll turn him over to you. Our Government refused. Being a guest of an Afghan tribal society is a really big deal in their culture. I think the Taliban's request for proof was reasonable.] The funny thing is, I've never heard the DC Plutocrats use any Treaty to justify the wars in question. Usually they claim the right to go to war based on the War Powers Act and/or <u>the duties of Commander-in-Chief</u>. They certainly couldn't use a Treaty to justify invading Iraq in 2003; Saddam hadn't attacked any other country then, whether our ally or not. The Bush Administration's justification was that Saddam was guilty of PRE-Crime. As Bush put it, "If we wait for the smoking gun, it will be too late."...sounds good, but that's a completely illegal reason to attack and invade another country.

As to <u>duties of the Commander-in-Chief</u> re protecting our country, he/she manages the war **after it has been declared by Congress**. The President is not allowed to start a war. If the crisis is a rebellion/insurrection caused by U.S. citizens, then yes, the military can be used. If foreign *criminals* (not soldiers) are attacking us, then it's a matter for Law Enforcement.

Any way you examine supposed justifications for the above-listed wars, they turn out to be illegal, unconstitutional, unethical, and even immoral. Welcome to the Soft Fascist States of America, a land I love, but sincerely regret the fact that we no longer have a representative government...and that's been true for decades. When he said we no longer have a representative government in this country, Professor Carroll Quigley was right in 1966.

It's a genuine shame.

October, 2014

Obama's Wars = NeoCon, Kabuki Theater

On *Democracy Now!*, Jeremy Scahill (author of the NYT bestseller, *Dirty Wars*) described Obama's drone wars as "Kabuki Theater". He implicitly referred to the Iraqi-Syrian ISIS War as the same. Mr. Scahill is right on the money. Let's also include the War in Afghanistan, the present purpose of which is unclear to just about everyone. Any group having anything to do with 9-11 has left Afghanistan long ago.

Almost three years ago, *Democracy Now!* ran a clip of General Wesley Clark relating a story of his visit to the Pentagon just after 9-11. https://www.youtube.com/watch?v=r8YtF76s-yM An old colleague of his, another General, pulled him into his office and said that we were going to invade Iraq. Clark asked "Why?", and his friend replied, "I don't know.". Sometime later, General Clark was back visiting his friend and learned that the NeoCons at the highest levels had plans to invade seven Muslim countries and institute regime change. Iraq, Syria, Libya, and Somalia were on that list. So was Iran. Such plans violate both the *U.S. Constitution* and International Law.

Enter Barack Obama, the Peace candidate in 2008 and the winner of the Nobel Peace Prize in 2009. The guy was going to protect whistleblowers, close Guantanamo, institute a transparent Administration, restore credibility to the Executive Branch, and end wars. This 2009 winner of the Nobel Peace Prize has bombed and/or launched missiles into Afghanistan, Iraq, Pakistan, Yemen, Libya, Somalia, and Syria. [You couldn't make up this stuff if you tried.] His war activities aren't that much different from those of the previous NeoCons...the guys who say he's a "socialist". Obama, the fellow who protects mega bankers from prosecution, is the best friend of the private defense industry, and expands war to the point where it's perpetual...supposedly is a socialist. Looks to me as though he's closer to being a NeoCon than anything else. He's a lover of Crony Capitalism. [*His vaunted Obamacare is nothing more or less than a **bailout** of mega health insurance companies. The "Leftist" Roosevelt Institute said exactly that in* *one of their analyses.*]

Meanwhile, the Constitution is being violated, war "authorizations" either nonexistent or long expired ["...the Iraq War Resolution..."] are used to wage Perpetual War, the meanings of key terms (such as "imminent" and "war") are totally changed, innocent people are being killed, and phony threats are manufactured. Jeremy Scahill thinks no one in Syria even has heard of the "Khorasan" Group. If it does exist, best estimates are that the very few members <u>are leaders of Al-Qaida</u>...hardly a NEW threat, and no more of an "imminent" danger than they've been for years.

Kabuki Theater for sure.

October, 2014

Obama's Latest War Propaganda

On *60 Minutes* this past Sunday, our President essentially said this (paraphrased) about the current ISIS war: it's not really a war; instead, it's counter-terrorism. The unmitigated gall of the man finally has surpassed that of Dubya Bush and Dick Cheney, something which I didn't think was possible. Mr. Obama, though I am loathe to lecture a President, I shall, simply because you appear to be in dire need of it.

You're not fooling us. We know full well that you recognize the following. When the military of one country crosses borders into other countries and then launches missiles and/or drops bombs, thus killing people and destroying infrastructure, *that's known as **war***. Your attempt to sidestep the issue of your violation of the *U.S. Constitution* by saying it's not really war is grossly pathetic. Incidentally, Mr. President, <u>the NYT's editorial board and a Law Professor who wrote a *Times* Op-Ed piece agree that you are violating the Supreme Law of the Land</u>. [The fact that the *Times* agrees is truly monumental.] Just so you know: we know that your latest war propaganda (it's not *really* a war) is BOGUS. The fact that you have a Constitutional Law background further convinces me that you, as well, know that it's bogus. [Anyone who needs more convincing might want to ask Glenn Greenwald (a former Constitutional Law attorney and present day Pulitzer-Prize-winning independent journalist) about it.]

As to Republicans at the highest levels in our Government, you should all be thoroughly ashamed of your lack of action relative to defending the *Constitution.* Oh, I'm sorry, I forgot: both branches (Democrats and Republicans) of the only major political party in the USA---the Mega Transnational Corporatist Party--- abandoned the Supreme Law of the Land decades ago. In spite of that (and because of that), you should all be thoroughly ashamed. You're all apparently putting the State-Corporate Complex ahead of We the People.

After WW II, when the name of the War Department was changed to the

Defense Department, someone (I forget who) remarked: that's Government-Speak for: from now on, all our wars will be aggressive (as opposed to defensive) and/or unconstitutional. That person was prophetically correct.

Lastly, as others have pointed out previously (with examples), no military can defeat a terrorist organization. History has proven that. [There may be an exception or two in the historical record.] Terrorists are criminals; terrorism is a Law Enforcement issue. In general, military personnel are not cops and shouldn't be expected to function as such. Recent illegal wars by the U.S. Fed Government serve the primary function of enriching huge, transnational corporations. That is so obvious that it's painful.

[Sidebar--- Steve Kroft of *60 Minutes,* the journalist who interviewed President Obama, is one of the very, very few Corporate Media employees who exposes the Corporatocracy. Kudos to Mr. Kroft!]

April, 2019

"War is a Racket", the title of a book by General Smedley Butler, first published in 1935

Going from the rank of Second Lieutenant to Major-General and after thirty-three years of service in the Marines, Butler said he realized he had been nothing but a "thug" for Big Corporations. His book is only forty-two pages long, but became an antiwar classic which still is brutally relevant today.

Only people who don't know what's going on still believe that our brave and well-meaning military folks are fighting for freedom, democracy, etc. In actual fact, they are the *unwitting* enforcement arm of our Fed Reserve Bank and other Western World Mega Banks, as well as other major Corporations whose leaders believe they are the real rulers of the world. Butler clearly pointed that out in his book, *War is a Racket*.

Now (and for some time) that both Repubs and Dems in DC pretty much ignore and/or violate key parts of the *U.S. Constitution*, the Corporatists are free to complete their establishment of Global Private Rule. Public Governments still will exist, but only as window dressing. By the way, none of this is a "grand conspiracy"; to the Powers-That-Be, it's simply "good business practice" and "the best thing for the world". It's also important to grasp the fact that **Corporatists are NOT Free Market Capitalists**; instead, they are Fascists. The very last thing they want is a Free Market. They prefer a series of Monopolies supported by their puppet governments.

To those who believe Trump is fighting all this, and is the champion of the Common Man, you are sadly mistaken. Trump is the quintessential elitist, Corporatist, Establishment, **Crony Capitalist**...the worst kind of capitalist. He's a serial liar, a Super-Rich snob, a gaslighter, and doesn't care at all about the bottom 90-99% of the USA's citizens. He's just telling you what he thinks you want to hear. His whole history proves all the above. Study it, and then you'll see. A good start is the documentary, "Meet the Trumps: from Immigrant to President"; it can be found online.

p.s. I'm an Army Vet and a gun owner. Not only Vets, but ALL of us need to wake up to the fact that War (especially the current and recent ones) is not about "spreading democracy". Instead, it's about economics and finance. The main beneficiaries of it are Mega Corporations, especially Mega Banks and so-called Defense Manufacturers/Contractors. Military personnel are being misused, and elitists such as Trump have little-to-no respect for them. With Trump, in particular, his own words prove that.

June, 2019

FINALLY – Someone in Congress Asks the Right Question

U.S. Congressional Representative, Ted Lieu of California, asked a Trump Administration official (in a Hearing) the following: "Under the Constitution, does the President have the authority to declare war?". The Trump guy hemmed, hawed, beat around the bush, and never answered the query. Then Rep. Lieu said he would make it easy: "The Constitution gives Congress the power to declare war, right?". More hemming and hawing from the U.S. Envoy to Iran. Kudos to Representative Lieu.

Those who rely on the War Powers Resolution (often called the War Powers Act), or any other "resolution", to declare war are violating the Constitution and their Oath of Office. In our system of government, it isn't up to the President to decide when to go to war. The Founders believed that could be subject to abuse, and so <u>they made certain that only Congress had that power</u>. At least then, We the People supposedly would be represented in the decision. Plus, Congress was not given the authority to transfer that power to the Executive. The President only has power to *manage* a war...once Congress has declared war. Finally, no Law (or congressional resolution) supercedes the U.S. Constitution.

According to recent News reports, Trump and Crew are considering using the 2001 Resolution as authority to go to war with Iran. That's ludicrous. Not only is it ludicrous (and illegal) because of the paragraph immediately above, but also because that Resolution had *absolutely nothing to do with Iran.* Nothing at all.

The last time Congress declared war was in December, 1941. Over the decades since then, members of our Gov't have ignored the war clause of Article I, Section 8 of the *Constitution of the United States.* That means each of them has violated his/her Oath of Office. Very few people seem to care.

The case for war with Iran is so flimsy that it really doesn't exist. The warmongers' efforts to make the case have a strong stink about them. More importantly, the President does not have the Constitutional authority to declare war on any nation. Once any part of our bedrock Supreme Law of the Land is by-passed, then the same can happen to any other part. <u>Hello</u>. This whole scenario reeks of "The Grand Chessboard: American Primacy and Its Geostrategic Imperatives", by Zbigniew Brzezinski. Empire on steroids.

It all demonstrates why citizens must remain vigilant regarding politics, and especially in relation to warmongering. Sorry to say, but after decades (since 1956) of keenly observing USA politics, I've come to the following conclusions---

- Most politicians desire war.
- War distracts the public from any economic problems.
- War is good for several sectors of Big Business, the same ones that make significant political re-election campaign contributions.
- War reinforces the toxic side of male energy, and that more or less assures continued warring.

There's a better way forward. We just need the will to pursue it.

October, 2023

Genocide in Gaza

https://www.democracynow.org/2023/10/26/
judith_butler_ceasefire_gaza_israel

Isn't the killing of over 7,000 civilians (almost 3,000 of them CHILDREN) enough? According to Israeli Leadership thinking, apparently not.

If you truly don't believe the actions of Israel are genocide, then perhaps you should listen carefully to the philosopher, Professor Butler, at the link above.

No one I've heard is justifying the brutal murders of Israeli civilians by Hamas. Those actions were criminality at its worst. On the other hand, the actions ordered by the Leadership of Israel in response are over-the-top, beyond-the-pale, immoral, and unethical. **Even many Jewish people**, in Israel, the USA, and other countries, are strongly protesting that response.

In Israel and the USA, government spokespersons essentially have stated: this is war, it's messy, and civilians get killed. In other words, too bad, such is life.

Sorry, that's not an ethically acceptable reason for the wholesale bombing of Gaza, which includes killing people in hospitals, schools, UN shelters, residences, and the like. Nor is it justifiable to bomb entire neighborhoods *because Hamas tunnels are under those neighborhoods*. Israeli Leaders seem to be assuming everyone living in such places was complicit in the construction and use of those tunnels. The Leaders also appear to be assuming that all Gazans support the brutal murder of Israeli civilians. Or, maybe they just think: hey, tough luck, but that's war.

Here's a recent hypothesis proposed (which I believe is probable): because some people in the Government of Israel seem to believe ALL Palestinians support terrorism, the current crisis is being used to make life in Gaza so unbearable that all the residents migrate out of the area, or alternatively, die from lack of food, water, medicine, and other supplies.

Israel claims it is targeting only Hamas. Come on, that doesn't hold water, not a drop. Why? In addition to the bombing, **Israel has laid siege to Gaza - no food, no water, no fuel for generators, no medical supplies,** etc. *A siege targets everyone*. Yes, they <u>finally</u> allowed numerous supply trucks to enter. For two million people, the amount of supplies was a pittance. Inadequate.

All civil people everywhere should be objecting to the actions of Israel in Gaza. Not only is the situation tragic and brutal, it's not ethically justifiable. Finally, how extremely ironic it is that Israel is committing genocide.

p.s. Here's another irony: the three Abrahamic religions ALL worship the God of Abraham, but often (in many cases) still can't be civil to one another. It makes one wonder: will humans ever "live and let live"?

December, 2023

Israeli Leadership & Hamas are EQUALLY guilty of not only war crimes, but also stupidity

Israel---

1) Attacking facilities such as hospitals, residences, refugee shelters, schools, U.N. SHELTERS - where not only adult civilians but also CHILDREN are present - is totally inexcusable. It's immoral, unethical, and unlawful. It's criminal, no matter what the excuse for it. [Have other countries (including the USA) done anything similar? Yes, *but such does not make it right or legal*.]

2) Telling refugees to evacuate to the south for safety, and then attacking the south, killing and maiming civilian adults AND CHILDREN, also is a war crime.

3) The Leadership is stupid if they think the CRIMINAL acts of Hamas justify "collective punishment" in Gaza AND THE WEST BANK. That doesn't fly in a civil society. They're also stupid if they believe ethnic cleansing is justifiable, and if they think the world doesn't know they are trying to get all Palestinians to leave both Gaza and the West Bank. Finally, they're stupid if they don't see that their actions are creating a whole new generation of militants.

Hamas---

a) Because of its scale and immoral brutality, their initial attack was not only criminal, but also a WAR CRIME. Period.

b) Hamas is stupid because they **had to know how egregious** the Israeli response would be. They had to know civilian "shields" wouldn't mean a damned thing to the Israeli Leadership. Why? Because the current Leadership has been strangling Gaza for quite awhile, and bulldozing Palestinian homes in the West Bank as well. They don't care one whit about the lives of Palestinians. [I'm referring only to Israel's current Leaders, not all Israelis.]

SIDE NOTE--- Most of the Fed Government here in the USA *should be completely ashamed of their involvement in this whole scenario,* and of their <u>weak, tepid</u> criticism of the KILLING OF OVER 5,000 **CHILDREN**. The greatest amount of shame on you. Such will not be forgotten for a long time.

[NOTE: by March, 2024, just over 14,000 children had been killed.]

January, 2024

Wars around the world currently are being run by thugs

In many regions, civil society is being egregiously damaged, essentially torn apart. Putin (in his defense) was backed into a corner by expansion of NATO to the East (and his border); then he reacted like a thug. With October 7, the Leadership of Hamas became criminal thugs. [Yes, Gaza and the West Bank have been strangled by Israel's politicians for decades; but that's no justifiable excuse to brutally murder civilians, including children.] Netanyahu and crew certainly were entitled to a response; but for weeks their over-reaction has become criminal thuggery. No time to summarize the various thugs in parts of Africa and even in South America.

While Biden is not a thug, he 's supporting the thuggery of Israeli Leadership by supplying weapons, ammo, and military intelligence against Gaza. So, in essence, he gives aid to:
a) the bombing and/or ground assault of medical facilities, ambulances, civilian residences, schools, UN shelters, refugee camps, and whole areas where Palestinians were sent because they supposedly were "safe" areas; and,
b) the killing of CHILDREN; and,
c) the ethnic cleansing and genocide of Palestinians; and,
d) the intentional starving of them as well.
While all that may not make him a thug, arguably it does place him on the border of being not only unlawful and heartless, but insane as well. [No, I'm not trained in clinical psychology. But a wise man once told me: "You don't have to be a master carpenter to recognize a shack.":]

This is all so Orwellian that it's truly mind-boggling. Why people keep electing Establishment thugs (for example, Netanyahu and the micro-thug, Trump), to office is beyond my comprehension. If not thugs, we elect dogmatic, stuck-in-the-Past, corporatist-supporting people with no vision and zero capability of thinking comprehensively (e. g., Bush, Biden, and again, Trump).

In this country, Elizabeth Warren should be elected President, or someone along the lines (back when) of the Republican Senator, Mark Hatfield, a truly honorable politician; or, someone like George McGovern, another honorable Senator. Instead, we get either wannabe Fascists or slumbering, corporatist drones.

C'est la vie

Politics

February, 2023

Rigged in the USA

I posted this piece (below: "A Fraud...") almost seven years ago. It remains highly relevant today. This situation is exactly why Presidential Debates in election years are rigged to the hilt. Blame the Republicrats, otherwise known as the Mega Corporation Party. It has two branches: the RINOs and the DINOs.

....

"A Fraud...Upon the American People"

I remember well the days when the League of Women Voters, a *totally* nonpartisan group, was in charge of moderating Presidential Debates in this country. That was in the 1970's and 1980's. At the end of the League's stint, the Republicans and Democrats rejected about eighty proposed questioners for the debate...as well as numerous tough questions. Later, the Republicrats came up with a nontransparent contract <u>sanitizing the debate</u>, saying who should be in the audience, and determining the exact questions that would be allowed. **The League resigned with this statement:** "<u>A fraud is being perpetrated upon the American People</u>.".

That's how we wound up with **a private corporation run by the Dems and Repubs,** the Commission on Presidential Debates, being in charge of the debates. The word "Commission" makes it sound as though it's some sort of public, representative, all-encompassing entity. It isn't...not by a long shot. This private company (funded by corporations & foundations) determines *who is allowed to debate, who will moderate the event, what questions will be allowed, and who is allowed in the audience*. This is **rigging** to the Nth degree.

At least the top three "Third Parties" should be allowed to participate in the upcoming debates...regardless of any percentage in polls. Why not? Isn't that democracy in action? If the Republicrats claim that there isn't enough time for that amount of participation, that's pure hogwash. Look at how many candidates were in the Primary Debates. Four, five, or even six candidates wouldn't hinder anything in the major Presidential Debates. *Quite the opposite: such a scenario would enhance the choices for the electorate.* The fact is: without being in those debates, a candidate has little to no chance to garner votes. That's one way the Republicrat monopoly eliminates its competition.

[See the first featured segment on today's broadcast of *Democracy Now,* Thursday, August 18, 2016. Jill Stein, Gary Johnson, and others have something important to say.]

A private corporation should not be allowed to have complete control over those debates. That's not democracy...that's Soft Fascism.

Okay, you're all excited about politics. In this Rigged System I can't imagine why, but good for you. Now, why not insist that Oligarchical control be relinquished, and instead, genuine participatory democracy be instituted? Why not? Habit? Break out of the box.

May, 2014

Lies, the Elites, the Corporate Media, Politics, and Class War

Recently on *Democracy Now!,* Glenn Greenwald (a major player on a journalistic team that just won a Pulitzer Prize for Public Service) characterized the Corporate Media in America as "impotent, neutered, and obsolete". I agree; however, it's important to recognize that they were created to be that way. Their major function, as mega corporations, is to distract the public and support the soft despotism (in America) and the fascist intervention (abroad) by the central government. A long time ago, Walter Lippmann---the "Dean of American Journalism" and a *so-called "Liberal"*---stated that the American public consisted of people who were "meddlesome outsiders" (relative to government) and made up the "bewildered herd". He also believed, along with Edward Bernays (the Father of Public Relations), that controlling the public mind was especially important in a democratic society, and that public "consent" must be "manufactured" by elites.

In 1975, *Crisis in Democracy* was published by the Trilateral Commission, an elite, "Liberal" think tank founded by David Rockefeller and Zbigniew Brzezinski. The book-length report, which was authored by elite *so-called "Liberals",* stated the proposition that America had a huge problem, a crisis--- ordinary people were organizing (which is anathema to Oligarchies) and demanding a place at the political table. These insurgents actually wanted to determine their own destinies and the policies of their servant government. According to the *"Liberal"* elite authors, the solution to the "crisis" was to somehow return what Lippmann called the "meddlesome outsiders" and "bewildered herd" to their proper state of apathy and obedience. Thus began another cyclic push by the Oligarchy (in general) to manufacture or engineer the consent of the public to allow things to be run by the "smart" people, the Meritocracy. The process continues right up to the present day. [The so-called "Conservatives" operate in the same manner.]

We still live in a relatively free society here in the USA. Because of that, the Oligarchy cannot resort to brute force (in most cases) in order to get us to toe the NeoImperialist, Corporatist line. They wouldn't get away with it. The method used by the State-Corporate Complex to gain our "consent" is Edward Bernays style Propaganda, which is instrumental in their strategy of Distract-Divide-Conquer.   [None of this is a "conspiracy". It's right out in the open, but one must pay attention in order to see it.]

The biggest threat to the Crony Capitalist, State-Corporate, State-Capitalist, Mega Corporate Oligarchy in America is not communists, or terrorists, or narco-kingpins, or any other boogie-man...the biggest threat is ordinary *people organizing around a just cause*. None of us can do much individually to obtain and secure a fair and just society; but with solidarity in a group, much can be accomplished peacefully. <u>That's what the Oligarchy fears most</u>.

Because of such fear, you will notice (if you're not totally distracted) that the elites and their Institutions will either completely ignore or almost instantly attack, smear, or marginalize any group (or individual, for that matter) which proposes or supports any egalitarian issue or project. Whistleblowers are criminalized, labor unions are crushed, environmental groups are ridiculed or marginalized, alternative media are ignored, workers protesting poverty-level wages are ignored or marginalized or fired, alternative energy proponents are ridiculed or ignored, health care reform advocates are deemed "socialists", constitutionalists are ignored, poor people in any organization protesting almost anything are deemed "unworthy", opponents of so-called "Globalization" (which really means mega corporate tyranny) are deemed "behind the times", and on & on.

Elite propaganda in America comes from a diverse array of sources - the highest levels of the Fed Government, "Conservative" groups, "Liberal" groups, the Corporate Media, various public Institutions, various private Institutions, think tanks, etc. It usually originates at the very highest levels, but no matter the source, the message almost always amounts to some variation of this: the State-Corporate Complex knows what is best for you and our country; you should leave it to us, no matter how unfair, unjust, or illegal it might seem. We know best. You should accept on its face anything we say. Trust us. Our best and brightest minds are handling it.

The major offenders are those at the highest levels of the Fed Government (no matter which Party holds majority power) and those at the highest levels of the Corporate Media. A close third would be those elites in the large, private political organizations supporting "Conservatives" or "Liberals", Republicans or Democrats. In fairness to the last group, I should say that maybe some of them promulgate propaganda without knowing that the Game is rigged. The two major political parties in America mean virtually nothing anymore...they're strictly meant to serve as distractions from the mega corporate agenda. Anyone with even half a political eye surely can see that they're merely two branches of the Mega Corporation Party.

Our Fed Government (and Mega Corporations) now, without probable cause or a proper warrant, spies on all of us. Our Government now spies on our allies, and even individual Heads of State (e.g., Merkel in Germany) who are our allies. In his new book, *Nowhere to Hide,* Glenn Greenwald reveals an NSA training document (from Snowden) which instructs government operatives basically to get it all, all the digital info it can from everyone, not just those suspected of terrorism. After all, *anyone in the world* possibly could be a "terrorist". These are not the actions of a moral, ethical, just government. Instead, this is Soft Fascism, or what de Tocqueville called Mild Despotism. As long as the People are distracted, divided, and thus conquered, the Government gets away with it; no brute force is involved. It's all unconstitutional.

Obama was at a WalMart the other day, praising that company in a speech. WalMart, the biggest employer in the USA, and the biggest corporate destroyer of the American economy (not counting the mega banks)...and also, in terms of how its workers are treated, possibly the shabbiest employer in the country. But that mega corporation fits right in with the elites' view of how Globalization is good for everyone. Republicans, Democrats, "Conservatives", "Liberals" (all at the highest levels) appear to take that as an axiom. They all also appear to support the WTO, the IMF, the World Bank, and "Free Trade" Agreements, all of which are not favorable to poor people, or even to Middle Class people. Meanwhile, the rich become the super-rich.

The lies promulgated by our Fed Government (again, no matter who is in power), the Corporate Media, and various public and private Institutions are so numerous and egregious as to boggle the mind. Propaganda does work if it's repeated enough times. Elites in supposedly democratic societies do rely extensively on propaganda to mold the public mind. I do see the beginnings of an awakening by Main Street to those axioms.

For the most part, all the above make politics in the USA a fraud perpetrated by Oligarchs/Plutocrats upon the citizens of this great Land.

July, 2018

The Plutocracy Continues in the USA

Below is my email reply to an organic farmer friend regarding his comment (in an email) on my last post here on this site. He and I agree on the Plutocracy, but differ as to the contributions to that by Repubs and Dems. He thinks the former are much more responsible than the latter. My reply below is edited somewhat.

....

I mostly agree with your comments. But, even though FDR's Populism was significant, he was not ideal. Even he admitted that sometimes he had to side with Big Biz. Plus, I agree with Chomsky's thoughts on Nixon---because Tricky Dick was an opportunist, and because he sensed the mood of the people in the early '70s (except for the Vietnam issue), he became---for the wrong reasons--- the *last of the New Deal politicians* (in a way).

I know it sounds outrageous, but think about it. Nixon signed into Law (he could have vetoed these): OSHA (Occupational Safety & Health Act/Administration, 1970), NEPA (National Environmental Policy Act, 1970), the precursor to the Clean Water Act (the Fed Water Pollution Control Act of 1972), the Clean Air Act (1970), the Marine Mammal Protection Act (1972), the Safe Drinking Water Act (1974, proposed by Nixon, signed into Law by Ford), the Endangered Species Act (1973), **and he created the EPA in 1970**, after signing NEPA.

Repub actions aren't all bad. Plus, though they do it less than Repubs, Dems *gerrymander* voting districts, too...and Dems *manipulate* Primaries, too... and Dems *suppress* Populist candidates, too. And until a few years ago, Dems in Congress also practiced **Insider Trading** on Wall Street. Dems also have an interventionist Foreign Policy, and Dems also often are warmongers and Empire Builders. And Dems also support questionable regimes such as Saudi Arabia and Honduras (never mind corrupt Netanyahu's Israel).

The whole system is rotten, not just the Repub side of the overall Plutocracy. **Those who keep blaming "the other side"** *are not helping to* *correct the situation. All they're doing is prolonging the squabble and distracting from the overall problem, which is Plutocracy-Oligarchy- Empire Building.* It's not about Republicans or Democrats; it's about Plutocrats, and they're found in both major Parties.

It all continues with the ultimate Plutocrats, Trump and His Crew. What few crumbs he has thrown to everyday people are vastly outweighed by his favoring of the Oligarchy. Plus, the deregulation blitz ultimately will harm us both financially and physically. The Don continues with his blatant lies, and apparently, some people believe every word he says.

On the Dem side, Schumer and His Crew continue to pretend to be Populists...when in fact, they are Plutocrats and/or puppets of the Plutocracy.

Then, too, we now have the Mega Banks and Mega Hedge Funds more empowered than ever before. A number of fairly well known and reputable people have stated in the past that if the public ever discovered how the Big Banks *really* operate, then there would be a revolution in the streets tomorrow. By the way, keep in mind that a revolution does NOT have to be violent. *A massive, national boycott of Mega Banks would do wonders toward restoring the Financial Sector to sanity*. We don't do it essentially for only one reason - convenience. Shame on all of us.

June, 2022

"...a clear and present danger to American democracy", and more

The title above is how retired Judge Michael Luttig described Donald J. Trump and his followers. The kicker is this: Judge Luttig is one of the most respected **conservative** judges in the USA.

https://www.theguardian.com/us-news/2022/jun/16/trump-clear-present-danger-to-us-democracy-conservative-judge-warns

This should mean all Main Street people <u>still</u> in the Trump Cult will re-evaluate their mistaken beliefs, but I'm not holding my breath. Cult members rarely ever are rational in any sense of the word.

It also should mean the hijacked Republican Party finally will come to its senses and dump Trump. Politics being as irrational as it is, again I'm not holding my breath. In general, the two major Parties of the current era appear to care little about honor, principles, ethics, the Law, and especially, the Constitution. Instead, their number one priority is re-election...*at any cost*.

The entire political scene in the USA is an embarrassment to the country. It's also a danger. Plus, it has caused most of the world to lose respect for us. Finally, it's not really much of a stretch to say our political scene **is an existential threat** not only to us, but to all life on Earth as well. Why? Simply put, the actions or *lack of action* on the following "fronts" are a threat to the existence of our species & all others: nuclear weapons and nuke waste; unprecedented ecological damage; societal disintegration; economic chaos; perpetual war; a belief in unlimited economic growth; corporatism; corporate globalization; and politics lacking common sense and ethics. We must get off that path if Life is to survive and thrive.

Ecological economics and the sustainability movement provide a viable alternative to the situation described above. Do an internet Search for those subjects. [2024 NOTE: So does my second book, *Choices and Change on the*

Path to a Sustainable Existence – Best Choices for Navigating Through a Monumental Crisis. It's available on Amazon and elsewhere.]

Trump followers justifiably are fed up with the Establishment; in my opinion (and that of many others), however, they are making a grave mistake by thinking DJT is the answer to our nation's ills. He's essentially a con man, a serial liar, an opportunist, and a plutocrat of the worst kind. He tells people what they want to hear; unfortunately, too many apparently believe his lies.

On top of all that, he's a danger to democracy, the working class, the biosphere, and the world in general. His ignorance and incompetence truly are beyond the pale. It would take an entire book to provide the details; but luckily, most informed people already are aware of them.

IMPORTANT NOTE: I'm not a Republican OR a Democrat. Instead, I'm an Independent.

March, 2024

Ten reasons why it's time for President Biden to gracefully retire

1. President Biden, you tepidly stated the Israeli response to the initial Hamas attack was now "over the top". It's much more than that; it's **criminal**. Yet, you keep sending weapons and ammo to Israel, thus *supporting the response which you say is too much, "over the top".* **Hello!**

2. Every country (including the USA) should be sanctioning Netanyahu and other members of the current Israeli Leadership who are responsible for purposely killing thousands of children and other civilians. <u>Those victims *are not collateral damage*</u>. It was **known** they were in the hospitals, UN shelters, refugee camps, ambulances, residences, etc. which were attacked. Netanyahu et al. are Trump-on-steroids. That's what you're supporting, Mr. President. Wake up.

3. According to Oil Change International, twenty countries (including the USA) are planning to expand oil production for ***twenty-six more years***. https://priceofoil.org/content/uploads/2023/09/OCI-Planet-Wreckers-Report.pdf

4. You tout your climate mitigation efforts, but they really are a mixed bag. In accordance with number 3. above, you are expanding drilling for fossil energy on public lands. Plus, the Inflation Reduction Act (constantly bragged about by you and others) favors oil companies as much as efforts toward renewable energy.

5. Your Administration continues to pursue Julian Assange. His "crime" was exposing US military misdeeds, some of which were war crimes. Your backing of Netanyahu means *you are complicit in war crimes*. That's according to many humanitarian scholars at the U.N., a few politicians here in the USA, a few universities around the world, and several governments. Wake up. You're aiding in the murder of CHILDREN.

6. Yes, you're better than Trump as a Leader, but not better than so many other *potential* candidates. Trump is so unAmerican and has

such a cult mentality, but *still* might win the election if someone other than you isn't running against him. You don't seem to know how to deal with his idiocy. Like many others, you let him get away with constantly *changing the subject whenever he's caught in a lie or otherwise backed into a corner.* He acts like a spoiled brat, and gets away with it. Anyone who knows the first thing about debating could show him up to be the whining child he is in no time at all. He's a gaslighter, a serial liar, a bully, and his knowledge of almost everything is pathetic. He's not an unknown.

7. Unlike a few politicians, you don't seem to understand ecological overshoot at all. Plus, you don't appear to grasp **any ecological principles** (not just overshoot). Like other anachronistic politicians, you apparently believe infinite economic growth - which, by the way, is exponential - is compatible with the health of a finite planet. You don't see such as an oxymoron.

8. Like Obama and Trump, **you're updating our nuclear weapons arsenal** instead of eliminating it. You seem to have bought into the idea that a "limited" nuclear war is possible without any significant consequences. Sheer ignorance.

9. Trump has zero statesmanship, but yours isn't much better. Putin warned the West about NATO expansion to his borders, and wanted talks about Ukraine. Your Administration avoided talks. Instead, you were quick to jump into a proxy war.

10. **Your outmoded foreign policy is a disaster**. It's predicated on the idea of the USA as the world's police force. It's stuck in a "unipolar" world, with the USA as the undisputed Head Honcho. It fails to recognize the emerging multipolar world, in which other major nations actually get to have some influence in their regions. See the Jeffrey Sachs interviews on *Democracy Now* for details.

The election still is eight months away. Time to change horses. There are plenty of potential candidates who have a much better chance than you, Mr. President, of defeating the poorest excuse for a politician in the history of our country. Bow out gracefully, and you'll see a flood of (okay, maybe only several) other candidates enter the race. The wannabe dictator will be defeated first verbally, and then at the ballot box. Hands down.

June, 2018

The Only Republicans and Democrats are on Main Street

They certainly aren't holding Office at the Fed level. Of course, there are a few exceptions to that statement, but very few. Those exceptions mostly all have been marginalized by the Plutocratic System.

The political "Right" and the "Left" on Main Street today are stuck in the Past. Today's politicians <u>at the highest levels</u> are not Repubs, Dems, Conservatives, or Liberals. Instead, they are Neoliberals/NeoConservatives/ Corporatists/Oligarchs/Plutocrats (or their puppets) who use those old labels to manipulate the public in order to divide and conquer. This is so obvious that it's painful.

The whole political scene here in the USA is a complete sham. The purpose of it is to facilitate the greatest transfer of wealth (from most of us) *in the history of Mankind* to the Upper Crust . **And it's continuing under Trump...big-time.** [Because he's a con man and a serial liar, he has managed to convince his followers he's for the common person.] None of it is a grand conspiracy. Rather, it's people with common interests working sometimes together, but mostly independent of one another, to attain mutual/common goals.

I've been closely following American politics since 1956 (1956 is not a typo). The pattern is clear: Edward Bernays style Propaganda has been (and is being) used by those in Power to mold the public mind, to move all of us toward the goals of the Plutocrats. The Powers-That-Be figured out many decades ago that, in a relatively Free society, brute force generally cannot be used to increase and guarantee their opulence. Rather, that has to be done by means of ubiquitous Propaganda.

The above thoughts/statements are not original with me. Over the decades, I've learned these ideas from many, many reputable people... *all across the political spectrum.* Most all of them have been marginalized to a great degree by those in Power. That has been accomplished by means of unrelenting Propaganda and Rules concocted by the Powers-That-Be, especially in the USA, but largely throughout the Western World as well. Since 2007, I've given example after example [on my main blog] of the use of Propaganda in this great Land.

Until we stop drinking the Bernays Style Propaganda Kool Aid supplied by the Plutocracy/Oligarchy (which adheres to neither Republican or Democratic philosophy), we have little to no chance of preventing the move to a NeoFeudalistic Society. If we don't wake up, NeoFeudalism will reign supreme.

November, 2018

Election = a Mixed Bag – Plus, World Corporate Empire

In what should have been a resounding, slam-dunk rejection of the corporatist, plutocratic, serial-lying, ignorant but clever, law-breaking Trump Admin (and any candidate supported by The Don), the Mid-Term election turned out to be a mixed bag. There are many reasons for that, but in this essay I'll focus on a chief reason.

[Technically, all the election results have not yet been tabulated. This is being written just before midnight on Tuesday, Mountain Standard Time. Nevertheless, I feel comfortable with the "mixed bag" characterization. Of course, both sides undoubtedly will claim victory on Wednesday. Once again, and for any new readers here, I'm not a Republican or a Democrat.]

Just as during the Bill Clinton Era, today's Democratic Party (at the highest levels) is run by plutocratic corporatists...or their puppets. After Clinton left office, the DNC re-took power from the Clintonite DLC (Dem Leadership Council, which heavily leaned toward Corporatism). Unfortunately for genuine Populists, the DNC quietly adopted most of the policies of the outgoing, corporatist DLC.

So, today (and for a number of years), the only difference between so-called Democrats and so-called Republicans is on *most social Hot-Button issues*--- such as gun control, abortion, gay rights, etc.---and maybe one or two non-Hot-Button, domestic issues. <u>When it comes to the economy, finance, war, empire, monetary policy, fiscal policy, corporatism, and foreign policy</u>, the two Parties (again, at the highest levels) are pretty much the same.

All of the above means that the Dems in control of their Party don't want to rock the Corporate Boat. That's why, in the lead-up to the election, we heard little to nothing about the following---

1. About 50% of working families in the USA are either in poverty, or so close to the brink of it that the difference doesn't matter. They are in the "Precariat Class", a term coined by the economist and professor, Guy

Standing. People in this Class are insecure relative to their material and psychological existence. Their lives, through little to no fault of their own, have become extremely precarious.

2. Of the remaining 50%, about 35% of them are doing okay, but they still are in the Precariat Class...at least in terms of economic and financial stability. Their jobs and lifestyle are no longer secure. They could lose it all virtually in a few weeks, or at most, a few months.

3. The economy is "booming" only for about the top 20% or so of income earners. Despite any piddly wage increases of late, wages/salaries still essentially are stagnant (after adjustment for price inflation). That's been true for about thirty years. In the same period, the Upper Crust has increased its income by 700% on average.

4. Our crushing National Debt (which has the potential of further ruining our economy) is not primarily due to social spending; rather, the main cause is massive Corporate Welfare (including military spending). Corporate welfare DWARFS all social spending combined. The "Nanny State" in the USA is for corporations, not everyday people.

5. Our foreign policy kowtows to Mega Corporations...especially those in the Financial Sector. Our Fed Government is belligerent on the world stage because that ultimately benefits the Big Boys in Biz. [See the book, *War is a Racket,* by General Smedley Butler, a Marine for 33 years.] Our foreign policy is further bankrupting our country. Do you really believe that we need 800 military bases around the world? Do you really believe that we *still* need to be in Afghanistan...after seventeen years? Do you really believe that sanctions on a country are not an act of war? Do you really believe that we need *more* expensive nukes? Our military budget already is larger than those of the next top seven countries combined (including China and Russia).

6. The goal of Mega Corps is to privatize everything...worldwide. At the highest levels in DC, politicians are helping them do that...or, at the very least, not opposing them. The monopolists of Mega Biz are moving us all

toward *NeoFeudalism*, and the pace rapidly is increasing. In the not-too-distant future, the Middle Class no longer will exist.

7. Finally, our foreign policy is crucial mainly because it drives domestic policy. Surveillance, War, Trade Wars, Sanctions, Empire...it all drains economic security from We the People. It all enriches the Upper Crust at our expense. It serves to funnel spending away from any safety net for the most vulnerable, and increasingly, from any economic security for everyone who isn't in the top tier of income earners.

The Big Boys and Girls at the DNC ignore most of the above. Even the small Progressive Wing of the Democratic Party often appears to not get the significance of foreign policy vis-a-vis the World Corporate Empire. The Democratic hierarchy is basically the same as the Republicans when it comes to foreign policy, sanctions, war, and Empire.

It all comes down to more wealth for the already wealthy...and the top Democrats don't want to rock the boat. So they mostly ignore Corporatism, and instead, concentrate on immigration or Russian "interference". No wonder they can't slam-dunk the rejection of Trump and Crew...and if anything needs slam-dunking, it's that.

[2024 NOTE: Relative to what's described in the above 2018 essay, things are very similar today. We really are ripe for a much-needed Third Party victory in this year's election, *but I'm not holding my breath*. People still believe such a win is impossible, which makes it a self-fulfilling prophecy, eh? Too bad for us. We need a change from the RepubliCRAT (aka, Demopublican) Mega Business Party.]

Spirituality

March, 2016

"God", Physics, and the Grand Illusion

Recently I watched a documentary on YouTube entitled, "What We Still Don't Know: Are We Real?". It's a piece on Physics and Cosmology…and quite fascinating. It relays the discovery of the Cosmological Constant, a concept in Physics (a Law of Nature) that suggested to theoretical physicists the distinct probability of the existence of a "Creator". [Google Cosmological Constant for details.]

This did not sit well with some physicists, so eventually they came up with the hypothesis known as the Multiverse. That allowed them to hypothesize a Super-Intelligence in another Universe which possibly could create a silicon Simulation of our Universe…thus eliminating the need for a "Creator". As crazy as it sounds, that hypothesis is somewhat plausible.

I left a Comment (modified a bit here) proposing another option, as follows. [This is not my usual Blog fare, but hey, a little variety in thought never hurt anyone… neither did a little stretching of the mind.]
………………………………..

Instead of an all-powerful "God" (the Creator) or a Super-Intelligence in another Universe that created a Simulation, perhaps the following is true. [This idea, or something very much similar, has been around for thousands of years in the philosophy of Buddhism.]

It could be that there is a Universal Mind or Universal Consciousness (the word "God" has too much baggage attached to it to be used in this scenario) which has always existed, is NOT all-powerful…and is the ONLY thing that does exist. What we perceive as the Universe merely is Universal Mind's imagination, or dream (of some sort), **or something else unknown but similar to those two possibilities**. In other words, what we perceive as "real" really is the ultimate Grand Illusion.

We easily can imagine that such a thing is possible simply by considering our own experience with our own dreams. When we are dreaming, everything in the dream seems real; only when we *awaken* do we find out the truth of the matter. Buddhists say that until you awaken to the true nature of existence and the Universe, you essentially will be frustrated, unhappy, and suffering. A Zen Master long ago said, "Once you realize the true nature of the Universe, the only thing left to do is to have a good belly laugh."*:)*

Such a scenario or possibility as described above includes or encompasses all religions, philosophies, and beliefs. Nothing is slighted or excluded... and none of it is "real". In such a case, perhaps we humans finally can cease squabbling with and killing each other. Instead, we all can have a good belly laugh and enjoy "life".

One reasonably might ask: what would be the point of "life" in such a case? In my estimation, the point would be that Universal Consciousness wants to experience everything: the good, the bad, the wonderful, the ugly, the beautiful, and all else in between.

April, 2024

More Thoughts, and a Personal Note

Spirituality is a crucial aspect of humanity. Formal, organized religion is a part of that, but such does not mean belonging to a "church" is required in order for a person to be "spiritual". Some spiritual people belong to various religious groups, some don't. In most countries, people are free to choose.

In any case, it's a safe bet that a truly spiritual person <u>should</u> possess at least most (or perhaps all) of these qualities: a belief in something beyond the physical dimension of existence; an ethical character; a moral character; a compassionate character; and a tolerance of people with views/beliefs other than one's own. I believe that's a reasonable expectation.

Taking the last characteristic above – tolerance – let's look at it in terms of today's political and social/cultural scene in the USA, and probably in many other developed countries. Keep in mind, what follows here is *in general*. I imagine there are exceptions.

Intolerance spans the entire political spectrum. "Think and act as we do, or we'll denigrate and harass you... *because we know what's best for everyone*." On the one hand (the political Right), "God told us"; and on the other hand (the political Left), social evolution told us. In either case: "We're right, and you're wrong. Period.".

Both the Right and the Left are "True Believers" in the worst sense, although, the Right is a tad worse about it than the Left. Not too much worse, though. One thing is clear to me after being a keen observer of American politics since Mr. Benson's American Gov't class at HPHS in 1956: both the Right and Left are intolerant of "the other side", and almost totally so. Of course, that's a general statement; I'm sure there are a few exceptions. Those exceptions

Political prejudice reigns supreme (in general) among people <u>on the Right and the Left</u>, which means each views "the other side" as ALL (or at least, an overwhelming majority) having the same characteristics as the few who are running the show on each side. As long as that continues, we will not have a genuinely civil society. Ever.

> The truth is, no group ever has been as homogeneous as what the Right and Left say the "other side" is today. I've known some super fine people in my life. About half were on the Right, and half were on the Left. They all were law-abiding and civil to each other. Some were more Purple than Red or Blue. People of today could take a lesson from that.

Finally, regarding tolerance:
- We're not saying anyone should be tolerant of criminals, but if someone is not breaking the law while doing something differently from the norm, such should be tolerated.
- Intolerance can be passive or aggressive. Passive intolerance is okay. It's merely exercising a preference. For example, if you're in a situation you can't abide, simply excuse yourself and leave...peacefully.
- Aggressive intolerance is not okay. Yelling at people, slurring them, threatening them, and/or doing violence to them are all things a spiritual person must

not do if one hopes to maintain spirituality in his/her life.

101

- Another form of aggressive intolerance (which can be peaceful in a way) is banning a person from giving a planned speech. Let them speak. If people don't want to hear it, they can walk out. Such would have a bigger impact on the speaker than banning the speech from the get-go.

Spiritual people can be forceful and forthright, but they also simultaneously are peaceful and compassionate. Are they always that way? No, of course not. Even a Buddhist Monk or a Pope can get frustrated or angry. Nevertheless, all people who believe they have spirituality strive to live up to such.

I was raised as a Christian, a Lutheran. In my twenties, I became an informal non-denominational Christian. In my thirties, I converted to an informal version of Zen Buddhism. Also in my thirties, I developed a strong set of EcoEthics.

Now, in my early eighties, I'm still a Zen adherent. No regrets at all about that. At each stage of my spiritual development, I have felt just as spiritual as at any other stage. That brings us to the next essay, as follows.

April, 2024

The One True God?

According to a common categorization set-up, there are eight *major* religions in the world: Hinduism, Islam, Christianity, Judaism, Buddhism, Jainism, Sikhism, and Zoroastrianism. There are many so-called minor religions, and perhaps a few of them should be bumped up to the "major" category.

Organized religion also is categorized as "monotheistic", "polytheistic", or somewhat of a combination of those two. Regarding the combination, my understanding of Hinduism is that while it has a number of Deities, the Head Honcho (so-to-speak) is "Brahman", the Creator. Of course, like all religions, there are different forms/sects of Hinduism; thus general statements about the belief are somewhat problematic. So, for example, the Vedanta form is philosophically nondualistic (like Buddhism) and religiously monotheistic.

Perhaps the best known of the monotheistic beliefs are the Abrahamic religions of Judaism, Christianity, and Islam. While they all worship the God of Abraham (God the Father), each of them has a religious construct different from the others. Christianity further muddies the waters by believing in the "Holy Trinity" – one God consisting of three different entities, the Father, Son, and Holy Spirit. [As a kid, when I navigated through Lutheran "catechism class", the third entity was called the "Holy Ghost".]

My point in relaying all the above is as follows. Some religions (the ones with a monotheistic belief) claim to follow "the one true God". Others (the polytheistic ones) say they have the one true belief. The primary basis for both those claims seems to be the "Holy" writings underpinning each Faith. Such writings are accepted as "the Word of God", or at the very least, true no matter who wrote them. I propose something different.

Before this *hypothetical* scenario is described, let's be sure we're all on the same level of understanding regarding the word "hypothesis". It's not "proof" of anything; it's not even a "theory". A proposed hypothesis can be based on observation and thought, but it still may be only a "hunch", an educated guess, or merely a wild idea. No proof is necessary to put forth the idea. If the hypothesis is *tested* over and over, and it holds up, only then can it be said to be a theory. It's tested further, and only then does it become accepted as "fact".

I propose this hypothesis: it's at least *possible* that <u>the beliefs of ALL religions are "true"</u>. That's especially the case if you believe in an ALL-POWERFUL God, one who can create anything, anytime. I have no idea whether or not such a situation is *probable*, but it most certainly is possible. When you die, perhaps whatever your Faith dictates is what happens to your spirit. Each group/"church" would experience something a bit (or a lot) different from what others experience. One might ask, how would that be possible? If your God is all-powerful, how would it NOT be possible?

Some might say: but such is not described anywhere in our Holy writings. Those people who believe in a **personal** Deity (as a Zen Buddhist, I don't) often say God has a Plan, and "We don't know God's Plan". Okay, so if you don't know the Plan for you, I assume it's not in your Faith's Holy writings either, just as the hypothesis above isn't. Who says your all-powerful, personal God would have transmitted everything which was to happen TO EVERYONE at death? Everyone, not just people of your particular Faith.

One might ask: what would be the reason for having such a scenario? One thing's for sure, I have no way of knowing that to any degree of certainty. A wild guess: it contributes to a desire of Universal Consciousness (what many refer to as "God") to experience all there is to experience.

BOTTOM LINE---

IF the proposed hypothesis is valid (and I'm not saying it is), then it would render all past, present, and future religious strife-wars- hatred as pointless to the Nth degree. Such would improve our existence by orders of magnitude. People would have no reason to squabble over who has the one true Faith. We would all have it.

I'll never understand why many people feel everyone has to believe whatever they believe. Especially regarding religion, and especially in the Abrahamic religions, too many humans are overly zealous in that regard. Some are fanatical, even to the point of violence against "the other side". Is that living a spiritual life?

Whatever happened to live and let live? Must we continue down the path of the dark side of tribalism? Each of us is capable of choosing a different way forward.

Final Thoughts

January, 2024

The Mother of All Feedback Loops re Human Society

A short while ago, I posted the following as a small part of a new book description---

"the complex feedback loop involving constant economic growth, which requires more energy, which then creates more complexity in society, which then requires more energy, which then creates even more complexity, etc. is ruining our habitat;".
This essay is an expanded view of that concept.

In our past (and ongoing) pursuit of infinite economic growth, we've created a System, a "Superorganism" according to the ecologist, Nate Hagens. This System is so large, so filled with **complexity**, and so out-of-control that a) it can't be fully described, and b) it can't be reined-in, and c) it's ruining the ecosphere. It's the ultimate Rat-Race. It's a primary cause of our ecological overshoot. It's a runaway machine with no one at the helm. It's literally killing us, not only with pollutants, but also with food insecurity, the stress of job insecurity, inadequate health care facilities, wars, inequality and inequity, declining democracy, and the like. About twenty percent (maybe less) of Earth's human population are managing to stay afloat in the ongoing energy binge; the rest of us are in a slow descent to subsistence living.

[**The twenty-six richest people on Earth** own as much wealth as HALF the entire world's population combined. In a civil society, that's an obscenity. It's most likely also why the Rich and Powerful (perhaps with a few exceptions) want the runaway System to continue. They appear to not be long-term, comprehensive, holistic thinkers.]

The ever-increasing *complexity* is causing significant problems in the arenas of: global supply chains, manufacturing, transportation, consumption, extraction of natural resources, generation and distribution of electricity, health care equity, income inequality, institutional functioning, and more. As society becomes more complex, the need for more and more energy increases.

Even a High School biology student could tell you with a high degree of certainty that such a scenario is unsustainable on a planet with finite resources. Unfortunately, it appears most (not all) of the Powers-That-Be **are relying on ideology, not science,** to navigate through this current Crisis. They appear to believe the benefits of our 150-year energy binge are the norm, but fail to see the Mother of All Feedback Loops, and the negative externalities of fossil fuel use. They also don't seem to grasp the looming minerals problem relative to our necessary transition to Green Energy. Never mind the soil crisis and the impending food supply shortage.

According to the mining engineer and geologist, Simon Michaux, in about only five years there will be serious problems regarding necessary minerals for Green energy. In about ten years, agriculture also will be having significant problems. It all revolves around energy.

If we don't voluntarily reduce our energy use, **we'll be forced to by lack** of cheap, easily available, and abundant fossil fuels AND **lack** of cheap, easily available and necessary minerals (about 70 of them) for renewables. Less energy use will lessen eco-overshoot by leaps and bounds. It also will slow down the Rat-Race by lessening societal complexity. This all can happen if we phase out of infinite economic GROWTH, and phase into economic development in a steady state economy. [See https://steadystate.org/

January, 2024

Humans, all other life on Earth, technology, and the future

 As we begin a new year, consider a physicist's view of what our economic superorganism is doing to our quality of life. The ecologist, Nate Hagens, interviews Thomas Murphy. I especially enjoyed his unique method of putting human history into perspective, **which is covered in the first 6 minutes**. Don't miss it; at least watch that. [You can spare 6 minutes:]

Do a YouTube Search for: "The Simple Story of Civilization with Tom Murphy, Frankly #22, Nate Hagens".

In terms of the geological time frame on Earth, we "modern" humans (*Homo sapiens*) haven't been in existence for very long -- about 200,000 years or so. Our ancestors started about 2.5 million years ago. In a recent written essay -- *The Simple Story of Civilization* -- Prof. Murphy compressed our full time (ancestors + "modern") on this planet into a 75-year span. Using that time frame for our total existence, he calculated the following:

a) a mere five years ago, *Homo sapiens* (modern human) appeared;
b) only fifteen weeks ago, agriculture (and "civilization") began -- that's about 10,000+/- years ago in the geologic time frame;
c) barely **four days ago**, the Age of Science began -- about 400 years ago in geologic time;
d) only **a day and a half ago**, the Age of Fossil Energy started -- around 150 years ago in normal geologic time;
e) finally, just **twelve hours ago**, global and rapid eco-devastation began -- that's about 50 years ago in normal time.
The petrochemical industry is largely responsible for e) above.

According to Dr. Murphy, several important points arise from the scenario just described.

1) Prior to the advent of agriculture and settlements/towns/cities
(civilization), we lived more or less in harmony with Nature.

2) For thousands of years after the emergence of civilization, we still had
some degree of harmony with our habitat.

3) Nevertheless, the arrival of civilization started a path of accumulation of
possessions, development of hierarchies in the extreme, and efforts
to **control Nature**.

4) Then, only 150 years ago, we harnessed a Master resource –
 relatively cheap, abundant, and easily transportable fossil energy – and we
basically went berserk on overconsumption and increased attempts to
control Nature.

5) Little thought was given to unintended consequences.

6) Our current civilization was not designed to be sustainable. For one thing,
the economic system the Powers-That-Be worship is constantly and infinitely
craving more energy, and more production, and more consumption. More,
more, and more into infinity. It's not sustainable ecologically,
thermodynamically, financially, or socially.

7) Plus, we've "doubled-down" on efforts to control Nature, something
which never will be fully attained. Meanwhile, we're damaging/destroying
the basis of our biophysical existence. Such a system is destined to fail, and
we're getting close to that point.

8) Because the system (a Superorganism) is so large, and essentially
uncontrolled, **the failure will be monumental**.

Prof. Murphy's BOTTOM LINE is this:
Our current path clearly is not sustainable. The spurt of growth and
expansion we've seen in the last 150 years is not the norm in human history,
is temporary, and is unsustainable to the max. Future choices must ALL be

predicated on sustainability principles which conserve and protect the ecosphere. In short, he's saying something I've echoed a number of times: **everything must change** if we are to survive and thrive. Only then will we be able to bequeath a healthy habitat and society to our descendants and their descendants.

How to bring about necessary changes is an open question. A good start would be to ignore Corporatocracy propaganda, shun dogma, open our minds, be more cooperative with others, live with intention, and increase our thinking out-of-the-box. Many people and nonprofit groups are doing such, and even a few parts of some governments are doing so. Best choices for moving forward are emerging, and should be supported. Make time for that. Little else is more important to our children and grandchildren.

A bit of wisdom from an old Inuit song (paraphrased)

From time to time, we all harbor some degree of concern or fear.
Sometimes it's on the surface, other times down deep. One thing
which lessens it is the coming of a new day and the sweet light of
dawn. Again we see what a beautiful and magnificent gift Earth is,
and our spirit rejoices.

Happy Trails, Scott Haley

p.s. If you found value in this book, please consider leaving a brief
comment/review on Amazon, or elsewhere. Thanks so muc

Recommended Reading

[All are excellent. The few in bold-faced type are especially so.]

Capra, Fritjof. 2002. *The Hidden Connections – Integrating the Biological, Cognitive, and Social Dimensions of Life Into a Science of Sustainability*

Commoner, Barry. 1971. *The Closing Circle – Nature, Man, and Technology*

Czech, Brian. 2021. *Supply Shock – Economic Growth at the Crossroads and the Steady State Solution*

Daly, Herman E. 1996. *Beyond Growth: The Economics of Sustainable Development*

Goudie, Andrew S. 2019. *Human Impact on the Natural Environment – Past, Present and Future – Eighth Edition*

Heinberg, Richard. 2011. *The End of Growth – Adapting to Our New Economic Reality*

Lent, Jeremy. 2021. *The Web of Meaning – Integrating Science and Traditional Wisdom to Find Our Place in the Universe*

Michaux, Simon P. March 2021. *Restructuring the Circular Economy into the Resource Balanced Economy*, Geological Survey of Finland

Orr, David W. 1994. *Earth in Mind – On Education, Environment, and the Human Prospect*

Ratcliffe, Jennie M. 2021. *Nothing Lowly in the Universe*

Rees, William E. June, 2015. *Economics vs. the Economy*, Great Transition Initiative [Publisher, online version]

Schumacher, E. F. 1973. *Small is Beautiful – Economics as if People Mattered*

Suzuki, David and McConnell, Amanda. 2002. *The Sacred Balance – Rediscovering Our Place in Nature*

Swartz, Brian and Mishler, Brent D. (editors) 2022. *Speciesism in Biology and Culture - How Human Exceptionalism is Pushing Planetary Boundaries*

Washington, Haydn (Edited by). 2020. *Ecological Economics – Solutions for the Future*

Watts, Alan. 1971. *Does It Matter? – Essays on Man's Relation to Materiality*

Wilson, Edward O. 2016. *Half-Earth: Our Planet's Fight for Life*

NOTE: The above list of Reading is <u>not a bibliography</u>. Essays don't always require such. Sources used for this book are embedded throughout the text.